Randomization in
Clinical Trials

Randomization in Clinical Trials

Theory and Practice

Second Edition

WILLIAM F. ROSENBERGER
George Mason University

JOHN M. LACHIN
The George Washington University

Published by John Wiley & Sons, Inc., Hoboken, New Jersey
Published simultaneously in Canada

For general information on our other products and services or for technical support, please contact our Customer Care Department within the United States at (800) 762-2974, outside the United States at (317) 572-3993 or fax (317) 572-4002.

Wiley also publishes its books in a variety of electronic formats. Some content that appears in print may not be available in electronic formats. For more information about Wiley products, visit our web site at www.wiley.com.

Library of Congress Cataloging-in-Publication Data:

Rosenberger, William F.
   Randomization in clinical trials : theory and practice / William F. Rosenberger, John M. Lachin.
      pages cm
   Includes bibliographical references and indexes.
   ISBN 978-1-118-74224-2 (cloth)
1. Clinical trials. 2. Sampling (Statistics) I. Lachin, John M., 1942- II. Title.
   R853.C55R677 2016
   610.72′4–dc23

                                                                      2015024059

Cover image courtesy of Getty/Shikhar Bhattarai
Typeset in 10/12pt TimesLTStd by SPi Global, Chennai, India

MIX
Paper from
responsible sources
FSC   FSC® C013604
www.fsc.org

Contents

Preface

PREFACE TO THE SECOND EDITION

Thirteen years have passed since the original publication of *Randomization in Clinical Trials: Theory and Practice*, and hundreds of papers on randomization have been published since that time. This second edition of the book attempts to update the first edition by incorporating the new methodology in many of these recent publications. Perhaps the most dramatic change is the deletion of Chapters 13–15, which describe asymptotic methods for the theory of randomization-based inference under a linear rank formulation. Instead, we have added several sections on Monte Carlo methods; modern computing has now made randomization-based inference quick and accurate. The reliance on the linear rank test formulation is now less important, as its primary interest was in the formulation of an asymptotic theory. We hope that Monte Carlo re-randomization tests will now become standard practice, since they are computationally convenient, assumption free, and tend to preserve type I error rates even under heterogeneity. The re-randomization techniques also reduce the burden on stratified analysis, and this description has now been folded into a single chapter on stratification, which is now separate from the chapter on covariate-adaptive randomization. Covariate-adaptive randomization, while still controversial, has seen the most growth in publications of any randomization technique, and it now merits its own chapter. We have also added a section on inference following covariate-adaptive randomization, along with (sometimes heated) philosophical arguments about the subject. Many new restricted and response-adaptive randomization procedures are now described. Many homework problems have been added.

Acknowledgments: This work was completed while W.F.R. was on a 1-year sabbatical during 2014–2015. In Fall 2014, he was supported on a Fulbright scholarship to visit RWTH Aachen University, Germany, and conversations with Prof. Ralf-Dieter Hilgers, Nicole Heussen, Miriam Tamm, David Schindler, and Diane Uschner were enormously helpful in preparing the second edition. In Spring 2015, he was Visiting Scholar in the Department of Mathematics, University of Southern California, and also at The EMMES Corporation. He thanks Prof. Jay Bartroff, Marian Ewell, and Anne Lindblad for facilitating these arrangements. He is also grateful to the many colleagues and former students who have helped him understand randomization better, in particular, Prof. Ale Baldi Antognini, Prof. Alessandra Giovagnoli, Prof. Feifang Hu, Prof. Anastasia Ivanova, Alex Sverdlov, Yevgen Tymofyeyev, and Lanju Zhang.

Alex Sverdlov was especially generous in his assistance on the second edition. He provided the authors with a bibliography of papers on randomization in the past 13 years and also suggested numerous homework problems that now appear in Chapter 9. Victoria Plamadeala developed Problem 6.11. Diane Uschner, Hui Shao, and Ionut Bebu assisted with some of the figures.

W. F. R.
Fairfax, VA

J. M. L.
Rockville, MD

PREFACE TO THE FIRST EDITION

The Department of Statistics at The George Washington University (GWU) was a hotbed of activity in randomization during the 1980s. L. J. Wei was on the faculty during the early 1980s and drew Bob Smythe into his randomization research with some interesting asymptotics problems. At the same time, John Lachin was working on his series of papers on randomization for *Controlled Clinical Trials* that appeared in 1988. He, too, was influenced by Wei and began advocating the use of Wei's urn design for clinical trials at The Biostatistics Center, which he directed at that time and now codirects. I studied at GWU from 1986 to 1992, taking many classes from Lachin, Smythe, and also the late biostatistician Sam Greenhouse. I wrote my doctoral thesis under the direction of Smythe, on asymptotic properties of randomization tests and response-adaptive randomization, topics covered in the latter chapters of this book. I also worked on clinical trials at The Biostatistics Center from 1990 to 1995 under the great clinical trialist Ray Bain (now at Merck). Needless to say, I was well indoctrinated in the importance of randomization to protect against biases and the importance of incorporating the particular randomization design into analyses.

Currently, I am continuing my research on randomization and adaptive designs at University of Maryland, Baltimore County, where I teach several graduate-level courses in biostatistics and serve as a biostatistician for clinical trials data and safety monitoring boards for the National Institutes of Health, the Veterans Administration and Industry. One of my graduate courses is the design of clinical trials, and much of this book is based on the notes from teaching that course.

The book fills a niche in graduate-level training in biostatistics, because it combines both the applied aspects of randomization in clinical trials along with a probabilistic treatment of properties of randomization. Although the former has been covered in many books (albeit sparsely at times), the latter has not. The book takes an unabashedly non-Bayesian and nonparametric approach to inference, focusing mainly on the linear rank test under a randomization model, with some added discussion on likelihood-based inference as it relates to sufficiency and ancillarity. The strong focus on randomization as a basis for inference is another unique aspect of the book.

Chapters 1–12 represent the primary focus of the book, while Chapters 13–15 present theoretical developments that will be interesting for Ph.D. students in statistics and those conducting theoretical research in randomization. The prerequisite for Chapters 1–12 is a course in probability and mathematical statistics at the advanced undergraduate level. The probability in those chapters is presented at the level of Sheldon Ross's *Introduction to Probability Models*, and a thorough knowledge of only the first three chapters of that book will allow the student to get through the text and problem sets of those chapters (with the exception of Section 3.5, which requires material on Markov chains from Chapter 4 of Ross). Chapters 13–15 require probability at the level of K. L. Chung's *A Course in Probability Theory*. Chapter 13 excerpts the main results needed in large-sample theory for Chapters 14 and 15.

Problem sets are given at the end of each chapter; some are short theoretical exercises, some are short computer simulations that can be done efficiently in SAS, and some are questions that require a lot of thinking on the part of students about ethics and statistical philosophy and are useful for inspiring discussion. I have found that students love to read some of the great discussion papers on such topics as randomization-based inference, the ECMO controversy, and ethical dilemmas in clinical trials. I try to have two or three debates during a semester's course, in which every student is asked to present and defend a viewpoint. Some students are amazed, for instance, that there is any question about appropriate techniques for inference, because they have been presented a single viewpoint in their mathematical statistical course, and have basically taken their instructor's lecture notes as established fact.

One wonderful side-benefit of teaching randomization is the opportunity to meld the concepts of conditional probability and stochastic processes into real-life applications. Too often probability is taught completely independently of applications, and applications are taught completely independently of probability and statistical theory. As each randomization sequence forms a stochastic process, exploring the properties of randomization is an exercise in exploring the properties of certain stochastic processes. I have used these randomization sequences as illustrations when teaching stochastic processes.

This book can be used as a textbook for a one-quarter or one-semester course in the design of clinical trials. In our one-semester course, I supplement this material with a unit on sequential monitoring of data. I assume that students already have a basic knowledge of survival analysis, including the log-rank family of tests and hazard functions. Computational problems can be done in SAS, or in any other programming language, such as MATLAB, but I anticipate students would be facile in SAS before taking such a course.

I also hope that this book will be quite useful for statisticians and clinical trialists working in the pharmaceutical industry. Based on my many conversations and collaborations with statisticians in industry and government, I believe the fairly new techniques of response-adaptive randomization are attractive to the industry and also to the Food and Drug Administration. This book will be the first clinical trials book to devote a substantial portion to these techniques. However, this book should not be construed as a book on "adaptive designs." Adaptive design has become a major subdiscipline of experimental design over the past two decades, and the breadth of this subdiscipline makes a book on the subject very difficult to write. In this book, we focus on adaptive designs only as they relate to the very narrow area of *randomized* clinical trials.

Finally, the reader will note many "holes" in the book, representing open problems. Many of these concern randomization-based inference for covariate-adaptive and response-adaptive randomization procedures, and also some for more standard restricted randomization, in areas of group sequential monitoring and large sample theory. I hope this book will be a catalyst for further research in these areas.

Acknowledgments: I am grateful for the help and comments of Boris Alemi, Steve Coad, Susan Groshen, Janis Hardwick, Karim Hirji, Kathleen Hoffman, Feifang Hu, Vince Melfi, Connie Page, Anindya Roy, Andrew Rukhin, Bob Smythe, and Thomas Wanner. Yaming Hang researched sections of Chapter 14 during a 1-year research assistantship. During the writing of this book, I was supported by generous grants from the National Institute of Diabetes and Digestive and Kidney Diseases and the National Cancer Institute. Large portions of the book were written during the first semester of my sabbatical spent at The EMMES Corporation, a clinical trials coordinating center in Rockville, MD. I am grateful to EMMES, in particular Ravinder Anand, Anne Lindblad, and Don Stablein, for their support of this research and their kindness in allowing me to use their office resources. On the second semester of my sabbatical, I was able to "test" a draft of the book while teaching Biostatistics 219 in the Department of Biostatistics, UCLA School of Public Health. I thank Bill Cumberland and Weng Kee Wong for arranging a visiting position there and the students of that course for finding a good number of errors.

W. F. R.
Baltimore, Maryland

I joined the Biostatistics Center of the George Washington University in 1973, 1 year after receiving my doctorate, to serve as the junior staff statistician for the National Institutes of Health (NIH)-funded multicenter National Cooperative Gallstone Study (NCGS). Jerry Cornfield and Larry Shaw were the Director and Codirector of the Biostatistics Center and the Principal Investigator and Coprincipal Investigator of the NCGS coordinating center. Among my initial responsibilities for the NCGS were to determine the sample size and to generate the randomization sequences. Since I had not been introduced to these concepts in graduate school, I started with a review of the literature that led to a continuing interest in both topics.

While Jerry Cornfield thought of many problems from a Bayesian perspective, in which randomization is ancillary, he thought that randomization was one of the central characteristics of a clinical trial. In fact, he once remarked that the failure of Bayesian theory to provide a statistical justification for randomization was a glaring defect. Thus in 1973–1974, Larry Shaw and I approached the development of randomization for the NCGS with great care. Larry and I agreed that we should employ a procedure as close to complete randomization (toss of a coin) as possible and decided to use a procedure that Larry had previously employed in trials he organized while a member of the Veterans Administration Cooperative Studies Program. That technique has since come to be known as the "big stick" procedure.

Later, around 1980, I served as the Principal Investigator for the statistical co-ordinating centers for the NIH-funded Lupus Nephritis Collaborative Study and the Diabetes Control and Complications Trial. Both were unmasked studies. In the late 1970s, I first met L. J. Wei while he was on sabbatical leave at the National Cancer Institute. He later joined the faculty at George Washington University, and we became close friends and colleagues. Thus, when it came time to plan the randomization for these two studies, I was drawn to Wei's urn design because of its many favorable properties. Later, I organized a workshop "The Role of Randomization in Clinical Trials" for the 1986 meeting of the Society for Clinical Trials. The papers from that workshop, coauthored with John Matts and Wei, were then published in *Controlled Clinical Trials* in 1988. During 1990–1991, I had a sabbatical leave, during which I began to organize material from these papers and other research into a book.

During 1991–1992, I taught a course on clinical trials in which I used the material from the draft chapters and my 1988 papers. One of the students auditing that course was Bill Rosenberger. Bill was concurrently writing his dissertation on large sample inference for a family of response-adaptive randomization procedures under the direction of Bob Smythe. Bob had conducted research with Wei and others on randomization-based inference for the family of urn designs. Bill went on to establish a strong record of research into the properties of response-adaptive randomization procedures.

In 1998, I again took sabbatical leave that I devoted to the writing of my 2000 textbook *Biostatistical Methods: The Assessment of Relative Risks*. During that time, Bill suggested that we collaborate to write a textbook on randomization. This book is the result.

In writing this text, we have tried to present the statistical theoretical foundation and properties of randomization procedures and also provide guidance for statistical practice in clinical trials. While the book deals largely with the theory of randomization, we summarize the practical significance of these results throughout, and some chapters are devoted to practical issues alone. Thus, we hope this textbook will be of use to those interested in the statistical theory of the topic, as well as its implementation.

Acknowledgments: I especially wish to thank L. J. Wei and Bob Smythe for their friendship and collaboration over the years and Naji Younes for assistance. I also wish to thank those many statisticians who worked with me to implement randomization procedures for clinical trials and the many physicians who collaborated in the conduct of these studies. Thank you for vesting the responsibility for these studies with me and for taking randomization as seriously as do I.

J. M. L.
Rockville, Maryland

1

Randomization and the Clinical Trial

1.1 INTRODUCTION

The goal of any scientific activity is the acquisition of new knowledge. In empirical scientific research, new knowledge or scientific results are generated by an investigation or study. The validity of any scientific results depends on the manner in which the data or observations are collected, that is, on the design and conduct of the study, as well as the manner in which the data are analyzed. Such considerations are often the areas of expertise of the statistician. Statistical analysis alone is not sufficient to provide scientific validity, because the quality of any information derived from a data analysis is principally determined by the quality of the data itself. Therefore, in the effort to acquire scientifically valid information, one must consider all aspects of a study: design, execution, and analysis.

This book is devoted to a time-tested design for the acquisition of scientifically valid information – the randomization of study units to receive one of the study treatments. One can trace the roots of the randomization principle to Sir R. A. Fisher (e.g., 1935), the founder of modern statistics, in the context of assigning "treatments" to blocks or plots of land in agricultural experiments. The principle of randomization is now a fundamental feature of the scientific method and is employed in many fields of empirical research. Much of the theoretical research into the principles and properties of randomization has been conducted in the domain of its application to *clinical trials*. A clinical trial is basically an experiment designed to evaluate the beneficial and adverse effects of a new medical treatment or intervention. In a clinical trial, often subjects sequentially enter a study and are randomized to one of two or more study treatments. Clinical trials in medicine differ in many respects from randomized experiments in other disciplines, and clinical trials in humans involve complex ethical issues, which are not encountered in other scientific experiments. The use of

Randomization in Clinical Trials: Theory and Practice, Second Edition.
William F. Rosenberger and John M. Lachin.
© 2016 John Wiley & Sons, Inc. Published 2016 by John Wiley & Sons, Inc.

randomization in clinical trials has not been without controversy, as we shall see, and statistical issues for randomized clinical trials can be very different from those in other types of studies. Thus this book shall address randomization in the context of clinical trials.

Randomization is an issue in each of the three components of a clinical trial: design, conduct, and analysis. This book will deal with all three elements; however, we will focus principally on the statistical aspects of randomization in the clinical trial, which are applied in the design and analysis phases. Other, more general, books are available on the proper conduct of clinical trials (see, e.g., Tygstrup, Lachin, and Juhl, 1982; Buyse, Staquet, and Sylvester, 1984; Pocock, 1984; Piantadosi, 2013; Friedman, Furberg, and DeMets, 2010; Chow and Liu, 2013; Matthews, 2006). These references also give a less detailed development of randomization.

1.2 CAUSATION AND ASSOCIATION

Empirical science consists of a body of three broad classes of knowledge: descriptions of phenomena in terms of observable characteristics of elements or events; descriptions of associations among phenomena; and, at the highest level, descriptions of causal relationships between phenomena. The various sciences can be distinguished by the degree to which each contains knowledge of the three classes. For example, physics and chemistry contain large bodies of knowledge on causal relationships. *Epidemiology*, the study of disease incidence, its risk factors, and its prevention, contains large bodies of knowledge on phenomenologic and associative relationships. Although a major goal of epidemiologists is to determine causative relationships, for example, causal relationships between risk factors and disease that can potentially lead to disease prevention, the leap from association to causation is a difficult one. Jerome Cornfield's (1959) treatise on "Principles of Research" gives a beautifully written account of the history of biomedical studies and the emergence of principles underlying epidemiological research.

Cornfield points to a mass inoculation against tuberculosis in Lübeck, Germany, in 1926. A ghastly episode occurred where 249 babies were accidentally inoculated with large numbers of virulent bacilli. In a follow-up of those babies, 76 had died, but 173 were still free of tuberculosis when observed 12 years later. If the tuberculosis bacilli cause tuberculosis, why didn't all the children develop the disease? The answer, of course, is the dramatic variability in human response to even large doses of a deadly agent. Thus, as we all know, tuberculosis bacilli cause tuberculosis, but *causation* in such cases, it does not mean that all those exposed to a pathogen will experience the ill effects.

Similarly, one can ask the famous question, why doesn't everyone who smokes develop lung cancer? One possible answer that would please the tobacco industry is that there is a hormonal imbalance that both causes lung cancer and causes an insatiable craving for cigarettes. An alternative answer is that there are *competing risks*: something else kills them first. The most probable answer is that not all those who smoke will develop cancer, due to biological or genetic variation.

Humans have such a complex and varied physiology; they are exposed to so many different environmental conditions; their health is also deeply tied to complex mental states. How can a scientist possibly sift through all the associations one can find between health and these other factors to find causes or cures for disease? One of the oldest principles of scientific investigation is that new information is obtained from a comparison of alternate states. Thus, a *controlled clinical trial* is an experiment designed to determine if a medical innovation (e.g., therapy, procedure, or intervention) alters the course of a disease by comparing the results of those undertaking the innovation with those of a group of subjects not undertaking the innovation.

Perhaps the first comparative study of record is the biblical account of Daniel (Chapter 1) in approximately 605 BCE, on the effects of a vegetarian diet on the health of Israelites. Rather than be placed on the royal diet of food and wine of the Babylonian court, Daniel requested that his people be placed on a diet of vegetables.

> "Test us for ten days," he said, "… then compare us with the young men who are eating the food of the royal court, and base your decision on how we look.…" When the time was up, they looked healthier and stronger than all those who had been eating the royal food.

Another famous example of a controlled intervention study is Lind's (1753) account of the effects of different elixirs on scurvy among British seamen. His study showed the beneficial effects of citrus and led (50 years after the study) to the Royal Navy's decision to store citrus on long voyages.

While the idea of comparing those on the innovative treatment with a control group sounds obvious to us today, historically it was not always entirely clear whom to include in the innovation and control groups. At the turn of the twentieth century, an antityphoid inoculation movement created controversy between Sir Almroth Wright, a famous immunologist, and Karl Pearson, who, along with Fisher, was a founder of modern statistics. Sir Wright gave the inoculation to anyone who wanted it and compared the subsequent incidence of typhoid with a group of men who refused the inoculation. Here is Pearson's first writing on the subject (Cornfield, 1959, pp. 244–245):

> Assuming that the inoculation is not more than a temporary inconvenience, it would seem possible to call for volunteers, but while keeping a register of all men who volunteered only to inoculate every second volunteer. In this way any spurious effect really resulting from a correlation between immunity and caution would be got rid of.

Four years later, Pearson's opinion was even stronger:

> Further the so-called controls cannot be considered true controls, until it is demonstrated that the men who are most anxious and particular about their own health, the men who are most likely to be cautious and run no risk, are not the very men who will volunteer to be inoculated.… Clearly what is needed is the inoculation of one half only of the volunteers, equal age incidence being maintained if we are to have a real control.

Pearson recognized what the immunologist did not: that human response to infectious, preventive, or therapeutic agents is variable and is positively related to patient characteristics, such as a willingness to volunteer to receive a new treatment. Thus, positive steps must be taken in the design and conduct of a study to eliminate sources of incomparability between those treated and the controls. The inoculated group cannot be compared to any arbitrary control group. The control group must be comparable to the treated group with respect to immune background, hygiene, age, and so on. Such factors are called *confounding variables*, because incomparability of the groups with respect to any such factors may confound the results and influence the answer to the research hypothesis.

These considerations play a major role in the design, conduct, and analysis of epidemiologic studies today. In an *observational epidemiologic study*, naturally occurring populations are studied to identify factors associated with some outcome. Since such studies do not employ a randomized design, the results are subject to various types of bias (*cf.* Breslow and Day, 1980, 1987; Rosenbaum, 2002; Selvin, 2004; Kelsey, *et al.*, 1996; among others). In a retrospective study, these populations consist of *cases* that develop the disease and controls that do not, so that a direct comparison can be made. Just as Pearson noted that there should be equal age incidence in both the inoculated and control groups, epidemiologists may also use *matching* on important variables (*covariates* or *prognostic factors*) that may confound the outcome. Matching is usually done, for instance, on important demographic factors, such as gender, age, and race. Each "case subject" will have a "control subject" with similar characteristics on matched covariates. This allows for greater comparability between the comparison groups. However, it is impossible to match on all known covariates that may influence outcome. Therefore, the leap from association to causation is again tenuous.

The most famous epidemiologic studies were those that demonstrated that smoking causes lung cancer. In 1964, the Report of the Advisory Committee to the Surgeon General was issued that led to warning labels on cigarette packages and restrictions on advertising. The report summarized the evidence from numerous studies that had shown an association between smoking and increased risk of lung cancer and other diseases. Despite any randomized controlled experiments, and based only on observational studies, the Committee concluded that the epidemiologic evidence showed that smoking was indeed a cause of lung cancer. The establishment of a causal relationship between tobacco smoking and cancer created much controversy (and does to this day in some circles). The Surgeon General's report on "The Health Consequences of Smoking" clarified the issue with a definitive statement on what types of evidence from observational studies can lead to a determination of a causal relationship. The Committee (1982, p. 17) stated:

> The causal significance of an association is a matter of judgment which goes beyond any statement of statistical probability (*sic*).... An entire body of data must exist to satisfy specific criteria;... when a scientific judgment is made that all plausible confounding variables have been considered, an association may be considered to be direct (causal)....

The Committee stated that the following five criteria must be satisfied:

1. *Consistency of the association.* Diverse methods of approach should provide similar conclusions. The association should be found in replicated experiments performed by different investigators, in different locations and situations, at different times, and using different study methods.
2. *Strength of the association.* Measures of association (e.g., relative risk, mortality ratio) should be large, indicating a strong relationship between the etiologic agent and the disease.
3. *Specificity of the association.* Specificity refers to the precision with which one component of an associated pair predicts the occurrence of the other component in the same individual. For instance, how precisely will smoking predict the occurrence of cancer in an individual? The researcher must consider that agents may be associated with multiple diseases and that diseases may have multiple causes. A single naturally occurring substance in the environment may cause the disease. A single factor can also be a vehicle for several different substances (e.g., tar and nicotine in tobacco), and these may have synergistic or antagonistic effects. There is also no reason to believe that one factor has the same relationship with a different disease with which it is associated. For example, smoking is also associated with heart disease, but perhaps in conjunction with dietary factors that are not important in lung cancer.
4. *Temporal relationship of the association.* Exposure to the etiologic agent must always precede the disease.
5. *Coherence of the association.* The association must make sense in light of our knowledge of the biology and natural history of the disease.

The nine largest studies cited in the Surgeon General's report comprised almost 2 million patients with 17.5 million patient-years of exposure. Based on these data and the convergence of evidence from other sources, one can be confident that smoking "causes" lung cancer, even though the precise causal agent has not been identified (i.e., tar, nicotine, or other agents) and even though no randomized experiment of the effects of smoking and lung cancer has ever been performed.

The overriding question in determining causality in such instances is whether the design or analysis has controlled or "adjusted" for all possible extraneous variables that might account for higher incidence of the disease. Some would say that only a randomized study can ensure adequate control for such factors. However, randomized studies of risk factors such as smoking are impossible, in that it is unethical to randomly assign risk factors to patients. It is instructive to note that Fisher, the father of randomization, was never convinced of the link between smoking and lung cancer, and perhaps equally instructive to note that he was a dedicated smoker. Today, almost all epidemiologists and biostatisticians will accept consistent, replicated, careful observational evidence, and few would argue the potency of the evidence against tobacco.

However, it is rare that an adequate body of evidence is amassed from epidemiologic studies alone to assert the aforementioned conditions. The number of studies, patients, and extent of exposure required to establish a definite cause by epidemiologic investigation are far greater, and the results ultimately less compelling, than those obtained from a randomized clinical trial, when such trials are possible.

1.3   RANDOMIZED CLINICAL TRIALS

In this book, we will refer to clinical trials that are prospective comparisons of two or more treatments, one or more of which is a new innovation under test, and one or more of which is a control. The most common is a *therapeutic trial*, in which a new therapy, such as a pharmaceutical agent (drug), is compared to a conventional therapy. In a *placebo-controlled* clinical trial of a new pharmaceutical agent, a group of drug-treated subjects may be compared to a group who receive a placebo control [a placebo being a drug preparation (e.g., pill) that is identical to the active therapy, but with inert (inactive) ingredients]. When an established therapy already exists, the new drug may be compared to an *active control*, where the control group receives the established therapy. Therapeutic pharmaceutical clinical trials are often called *phase III* clinical trials, because they represent the third phase of a four-phase process in investigating a promising new therapy. From development of a new pharmaceutical agent to its approval, there is often a *phase I* clinical trial, a small trial to determine the potential toxicity of different dose levels of a drug, and a *phase II* clinical trial, a preliminary study of toxicity and efficacy. A *phase IV* clinical trial involves post-approval follow-up of patient status. These phases are particularly seen in the study of cancer chemotherapeutic agents (see Buyse, Staquet, and Sylvester, 1984), and the four-phase process is often streamlined in other specializations of medicine.

The innovation, however, need not be a simple drug. In some cases, a new procedure is evaluated. An example is the Lupus Nephritis Collaborative Study that desired to assess the effects of plasmapheresis on the progression of lupus nephritis or kidney disease associated with lupus erythematosis (Lewis, *et al.*, 1992). Patients in the plasmapheresis group were hospitalized for a month to undergo daily plasma filtration and exchange, followed by the initiation of standard immunosuppressive therapy consisting of cytoxan and prednisone, the dose of the latter tapered when the patient responded favorably. The patients in the control group received comparable immunosuppressive therapy without initial plasmapheresis.

In other cases, a new intervention or an entire treatment regimen of multiple therapies is compared to a control regimen. An example is the Diabetes Control and Complications Trial, which was designed to assess whether a program of intensive therapy aimed at maintaining near-normal levels of blood glucose in subjects with type I diabetes mellitus would prevent or retard the progression of microvascular complications associated with diabetes. Patients in the intensive treatment group received aggressive insulin therapy with frequent monitoring of glucose levels, in conjunction with dietary counseling and exercise. Patients in the conventional treatment group received conventional therapy aimed at maintaining general well-being. While intensive therapy greatly reduced the risks of complications compared to conventional therapy, such an overall comparison alone cannot identify the mechanism by which the treatment had its effects (Diabetes Control and Complications Trial Research Group, 1993). Subsequent analyses, however, indicated that the effects of intensive treatment were indeed wholly accounted for by the reductions in blood glucose levels.

Some call such trials *pragmatic trials* because the innovation consists of two or more possible agents or procedures used in combination, such that the overall group comparisons alone cannot identify the mechanism by which the innovation produces its effects. However, the pragmatist would argue that conclusive evidence that the innovation is indeed beneficial in practice is adequate for its adoption even when the mechanism of the effect is unknown.

The pivotal component of phase III clinical trials is *randomization* or random assignment of patients to receive either the experimental treatment(s) or control. Cornfield (1959, p. 245) summarized the importance of randomization:

> 1. It controls the probability that the treated and control groups differ more than a calculable amount in their exposure to disease, in immune history, or with respect to any other variable, known or unknown to the experimenter, that may have a bearing on the outcome of the trial. This calculable difference tends to zero as the size of the two groups increase.
>
> 2. It makes possible, at the end of the trial, the answer to the question "In how many experiments could a difference of this magnitude have arisen by chance alone if the treatment truly has no effect?" It may seem mysterious that a mathematician could actually predict the course of future experiments. All you have to do is compute what would happen if a given set of numbers were randomly allocated in all possible ways between the two groups. Randomization allows this.

The first property of randomization is that it promotes comparability among the study groups. Such comparability can only be attempted in observational studies by adjusting for or matching on *known* covariates, with no guarantee or assurance, even asymptotically, of control for other covariates. Randomization, however, extends a high probability of comparability with respect to unknown important covariates as well. The second property is that the act of randomization provides a probabilistic basis for an inference from the observed results when considered in reference to all possible results. This randomization approach to inference is very different from the usual testing of unknown parameters arising from an independent and identically distributed sample from a known distribution. Later we will deal in detail with these and other precise statistical properties of randomization.

In Cornfield's first point, we come to the root importance of the randomized clinical trial. As scientists are interested in descriptions of phenomena, association among phenomena, and then mechanisms of causation, then the biomedical studies for each require increasing standards of evidence. Basic science research often involves the description of phenomena, observational studies lead to the determination of associations among phenomena, and randomized clinical trials lead to definitive statements on causative effects of agents or regimens on disease processes. As we have seen, despite the fact that consistent, replicated observational studies can also lead us to determine causality, there may always be questions as to whether we have controlled for all factors relating to incidence and prognosis of a disease. The randomized clinical trial allows this control and, hence, represents the highest standard of evidence among biomedical studies.

Among the first clinical trials, as we know them today, were the trials performed under the direction of Sir Bradford Hill in the 1940s by the Medical Research Council. These were the first medical trials to employ randomization to individual patients and constituted a major advance. They led to important findings in many of the persistent diseases of the day, such as whooping cough and tuberculosis. In every respect, they were similar to the most rigorous trials conducted today.

The polio vaccine trial of 1954 changed the face of public health worldwide (see Francis, 1955). Approximately 400,000 children were randomized to receive either the vaccine or a saltwater injection. The results showed a relative risk of 2.5, in favor of the vaccine group. The success of this study belies the controversy among study participants about the need for a controlled, randomized study. In fact, in a quotation attributed to Jonas Salk, it appears that Salk was not convinced of the need for a placebo control in the polio trial (source unknown):

> In talks with many people in our own group … and others as well, I found but one person who rigidly adhered to the idea of a placebo control and he is a bio-statistician who, if he did not adhere to this view, would have had to admit his own purposelessness in life.

In the end, randomized controls were felt necessary because of the variability of incidence of polio from year to year. It was largely due to trials like the polio vaccine trial that convinced the medical community of the value of the randomized clinical trial. Today, it is often considered the "gold standard" among techniques of ascertaining medical evidence.

A good example of the benefits of randomization can be seen in the National Cancer Institute's clinical trial of 62,000 women covered by the health insurance plan (HIP) of Greater New York, commonly known as the HIP trial (see Cairns, 1985). The women were randomized into a "test" group, who were offered a free annual physical examination and mammography for early detection of breast cancer and a "control" group who were given no special encouragement to be examined. The trial was designed to determine if the act of offering free mammography examinations reduces deaths from breast cancer. The results were encouraging. Among the test group, there were 2.9 deaths per 1,000 women in the first 9 years, and among the control group, there were 4.1 deaths per 1,000 women. The two groups were comparable in their incidence of breast cancer and in terms of general mortality from causes other than cancer, as should be the case because the experiment was randomized. But the results of the trial were also interesting because, among the test group, those who refused examination had a lower death rate due to breast cancer (2.8 per 1,000) than those who accepted the mammography (3.0 per 1,000). This demonstrates the danger of accepting observational data at face value, as one might have concluded that mammography was not effective. The acceptance and rejection groups within the test group were self-selected and, hence, subject to confounding due to incomparability with respect to important covariates. In this case, Cairns (1985) believes the confounding variable to be education level. Since better-educated women are known to be more likely to have breast cancer, and less well-educated women are more likely to have less interest in their health, and consequently are more likely to reject examination, the observational component of this study was biased in favor of the rejection group (for an instructive set of homework problems on the HIP data, see Freedman, Pisani, and Purves (1998), Problems 9 and 10, pp. 22–23.)

1.4 ETHICS OF RANDOMIZATION

Randomized clinical trials use probability as a method of assigning treatments to patients. Many have argued that probability has no role in medicine and that only a physician can decide which treatment a patient should receive, using his or her best judgment. However, clinical trials present a unique situation in which new innovations, such as investigational drugs, are being tested for efficacy and safety. Until a drug is proven to be effective and adequately safe, or ineffective or harmful, or just ineffective, the physician is in a state of *equipoise*: a state of genuine uncertainty about which experimental therapy is more effective. Most ethicists would agree, in principle, with the concept that it is ethical to employ randomization in a state of true equipoise, provided the patient consents to be a study participant and is fully informed about the potential benefits and risks of the treatments to be compared in the study.

However, ethics involving human experimentation are seldom so simplistic. On the one hand, a clinical trial gives the patient a chance of being assigned to a potentially beneficial therapy that would not be obtainable elsewhere. But that therapy may also be highly toxic. There is also a chance that a patient will be assigned to a placebo, in effect being denied a therapy that may later prove to be very beneficial (or, on the other hand, harmful). Decisions to enroll in a clinical trial are difficult ones, for this reason, and the patient must often be willing to make a sacrifice for the benefit of our public health.

These considerations exemplify the delicate balance between *individual ethics* and *collective ethics* (see Palmer and Rosenberger, 1999). Individual ethics dictate what is best for the individual patient, while in collective ethics, we consider the advancement of public health through careful scientific experimentation. In a broad sense, collective ethics leads to individual ethics, as it is only when careful scientific experimentation has yielded a universal standard of care for a given disorder that physicians will be fully informed and will have a scientific basis for the assignment of the best therapy to an individual patient. Although experimentation may lead to many patients being assigned an inferior therapy prior to the determination of the standard of care, this is the price an informed society must pay to obtain the evidence necessary to support informed therapeutic decisions. Such ethical dilemmas are naturally controversial and are the subject of many treatises and texts (e.g., Engelhardt, 1996).

Some would argue that equipoise is rarely present at the beginning of a phase III clinical trial. Animal studies and phases I and II clinical trials data, plus information on the biological action of the innovation (e.g., drug), combine to create in the mind of many physicians a belief in the effectiveness of one therapy over another. But such confidence may often be premature. The literature is replete with results of negative studies, where promising therapies were shown to be ineffective or even terribly harmful. If equipoise is defined in the confines of a single physician's "hunches" or intuition about a therapy rather than in a global standard of evidence based on randomized controlled studies, there will be no advancement of medical science. This is not to say that careful, replicated, consistent observational studies, as defined in Section 1.2, are not useful and cannot be convincing. But randomization adds an additional component that mitigates contention, and the National Institutes of Health and U.S. Food and Drug Administration now consider a well-conducted, randomized clinical trial to be of vital importance in demonstrating the efficacy and safety of a new therapy. As Cassel (2009) points out in the *Proceedings of the National Academy of Sciences*,

> While the ideal of ethics posits equipoise, the reality is a struggle in the minds of both the investigators and the patients themselves. The investigators are told not to posit the "new" treatment as "better," and yet the patients coming into the trial often agree to the trial precisely because they hope to find a better treatment for their condition.... The policy prescription for this problem was the requirement for informed consent, with clear guidelines about what kinds and levels of information were to be provided and what attributes of setting and patient condition constituted adequate freedom from undue influence for valid consent. Federal regulations established policy for federally funded research, and a substantial and costly infrastructure of internal review boards [(IRBs)] was established, both academic and free-standing. The scope of this ethics enterprise grew as the scope of the clinical trials industry grew, to the point where there are 4000 IRBs registered with the federal government... and > 26, 000 clinical trials were published in 2008 alone....

Some have also argued that randomized controls are unnecessary and unethical in studies where there are some data already available on the natural history and progression of the disease studied. Rather, they propose that a current cohort of experimentally treated patients might just as well be compared with a past cohort of patients receiving an earlier or no treatment, that is, a cohort of *historical controls*. In cases where one observes a complete dramatic reversal of the course of a disease, such as the effects of penicillin on a bacterial infection, such evidence may be convincing. However, most therapies yield modest effects and historical controls are subject to various biases that may easily skew the study results. The basic problem is that the historical control group might have very different characteristics from experimental cohort that may bias the study. Such factors might include patient selection criteria, diagnostic methods, the nature of follow-up observations, the criteria for response, and the extent of administration of concomitant medications. A difference between groups in any one of these factors or other factors could result in differences between groups with respect to study outcomes.

While most of today's scientists have embraced the randomized clinical trial, occasionally particular clinical trials arise that elicit passionate opposition on ethical grounds. A prime example is the recent clinical trials program in third-world countries on the benefits of short-term zidovudine (AZT) therapy in reducing maternal–infant HIV transmission. In a landmark clinical trial, Connor *et al.* (1994) show that 6 weeks of AZT therapy in pregnant women with HIV reduced the transmission to the infant by two-thirds. The results of this trial were hailed in the medical community, and 6 weeks of antiretroviral therapy quickly became the standard of care for HIV-positive pregnant women in the United States. Unfortunately, the prohibitive cost of zidovudine has prevented developing countries from implementing what is now the standard regimen in the United States. Consequently, a large group of scientists determined that clinical trials should be conducted in these countries using a shorter, less costly regimen of antiretroviral therapy, and such trials were begun with funding from the U.S. government. In an editorial, Lurie and Wolfe (1997) argue that placebo-controlled trials in developing countries are unacceptable, since an effective therapy had already been found in the United States:

> … On the basis of the [Connor *et al.* data], knowledge about the timing of peri-natal transmission, and pharmacokinetic data, the researchers should have had every reason to believe that well-designed shorter regimens would be more effective than placebo. These findings seriously disturb the equipoise … necessary to justify a placebo-controlled trial on ethical grounds.

In addition, they argue that, since the standard of care in developing countries (i.e., not providing therapy) is not based on consideration of alternate treatments or clinical data, and rather is based on economic considerations, researchers have an ethical responsibility to provide treatment that conforms with the standard of care in the sponsoring country (i.e., the United States).

This editorial led to much debate in the medical literature. Several of the researchers on these clinical trials in developing countries responded with their own editorial (Halsey *et al.*, 1997). They argue that a placebo control arm is necessary in order to determine if the short course of zidovudine is effective in these countries. Furthermore, they state that providing the same level of care routinely provided to mothers and their infants in the United States would violate the guideline to avoid undue inducements for participation in research and would make the research totally impractical.

> If these unsustainable services were provided on a temporary basis what would happen when the research project ended and local practitioners could no longer provide diagnostic tests, infant monitoring, and intensive care units necessary to support the regimen?

They close by noting that many dramatic interventions in developing countries could have been prevented had such "medical and ethical imperialism" been imposed on participants in international studies.

The Declaration of Helsinki was revised in October 2000, adding the following statement:

> The benefits, risks, burdens and effectiveness of a new method should be tested against those of the best current prophylactic, diagnostic, and therapeutic methods.

Although this does not exclude the use of placebo in studies where no proven prophylactic, diagnostic, or therapeutic method exists, this new directive is very controversial since some interpret it to mean that a placebo should never be used whenever effective therapy is available.

At issue, however, is not the act of randomization but rather the choice of the control treatment, either placebo or an active control (when the latter exists). Randomization of treatments to patients is now considered the seminal element of a clinical trial for the evaluation of a new innovation in medical care. The purpose of this book is to describe the theoretical basis for the various types or approaches to randomization commonly employed, to describe their statistical properties, and to describe considerations in their practical implementation.

1.5 PROBLEMS

1.1 From a recent issue of any major medical journal (e.g., *New England Journal of Medicine, Journal of the American Medical Association*), select an article that presents results of a controlled clinical trial involving at least 50 patients. The study should focus on a clinical result (i.e., effectiveness or safety of a treatment) rather than physiologic results (e.g., laboratory or physical measurements).

(i) Give a detailed description of the study design.

(ii) Provide a critique of the study design in regard to the potential for bias in the study results or conclusions. Did the authors describe the choice of study design well and describe possible pitfalls of the design?

(iii) Based on this study, if you were the statistician for a new study (either for a new treatment for the same disease or a study confirming results of the study), describe how you would design a study using randomized controls.

(iv) Alternatively, describe how you would design a study using nonrandomized controls.

(v) Discuss the implications for a randomized versus a nonrandomized study on the interpretation of the results. Which would be preferable?

1.2 From a recent issue of a medical or epidemiologic journal, select an article that presents the results of a nonrandomized observational study of a risk factor associated with an increase or decrease in the risk of a disease or adverse disease outcome.

(i) Give a detailed description of the study design.

(ii) Provide a critique of the study design in regard to the potential for bias in the study results or conclusions. Did the authors describe the choice of study design well and describe possible pitfalls of the design? Which possible biases are cited by the authors and what steps were taken, if any, to address them? Can you identify other possible sources of bias?

(iii) Based on this study, if you were the statistician for a new study (for either a new treatment for the same disease or a study confirming results of the study), describe how you would design a study using randomized controls, if possible.

(iv) Alternatively, describe how you would design a study using nonrandomized controls.

1.3 If you were the statistician on a steering committee, which is deciding whether to participate in a placebo-controlled clinical trial of maternal–infant HIV transmission and short-term AZT in a developing country, where the country has no access to the standard-of-care therapy in the United States (i.e., long-term AZT therapy), what would your stance be? Prepare a 5 minute position paper for a classroom debate. You are asked to respond to the following questions:

(i) Are such trials necessary and ethical?
(ii) Should any placebo-controlled study be adopted?
(iii) Are studies with historical controls reasonable?
(iv) What are the alternatives?

1.4 Are the considerations of individual and collective ethics the same in all clinical trials? Suppose you cross-classified a disease with respect to severity and incidence. For instance, you could have a 4-by-4 table with ordinal categories ranging from 1 to 4. For severity, the categories could range from 1 = mild to 4 = life-threatening. Similarly, incidence could range from 1 = very rare to 4 = very common. Within each cell of the cross-classification, determine the relative importance of individual versus collective ethics. (Palmer and Rosenberger, 1999)

1.6 REFERENCES

BRESLOW, N. E. AND DAY, N. E. (1980). *Statistical Methods in Cancer Research. Volume I - The Analysis of Case-Control Studies*. International Agency for Research on Cancer, Lyon.

BRESLOW, N. E. AND DAY, N. E. (1987). *Statistical Methods in Cancer Research. Volume II - The Design and Analysis of Cohort Studies*. International Agency for Research on Cancer, Lyon.

BUYSE, M. E., STAQUET, M. J., AND SYLVESTER, R. J. (1984). *Cancer Clinical Trials: Methods and Practice*. Oxford Medical Publications, New York.

CAIRNS, J. (1985). The treatment of diseases and the war against cancer. *Scientific American* **253** 51–60.

CASSEL, C. K. (2009). Statistics and ethics: models for strengthening protection of human subjects in clinical research. *Proceedings of the National Academy of Sciences of the United States of America* **106**, 22037–22038.

CHOW, S.-C. AND LIU, J.-P. (2013). *Design and Analysis of Clinical Trials: Concepts and Methodologies.* John Wiley & Sons, Inc., New York.

CONNOR, E. M., SPERLING, R. S., GELBER, R., KISELEV, P., SCOTT, G., O'SULLIVAN, M. J., VANDYKE, R., BEY, M., SHEARER, W., JACOBSEN, R. L., JIMINEZ, E., O'NEILL, E., BAZIN, B., DELFRAISSY, J.-F., CULNANE, M., COOMBS, R., ELKINS, M., MOYE, J., STRATTON, P., BALSLEY, J., and for the Pediatric AIDS Clinical Trials Group Protocol 076 Study Group. (1994). Reduction of maternal-infant transmission of human immunodeficiency virus type 1 with zidovudine treatment. *New England Journal of Medicine* **331** 1173–1184.

CORNFIELD, J. (1959). Principles of research. *American Journal of Mental Deficiency* **64** 240–252.

Diabetes Control and Complications Trial Research Group. (1993). The effect of intensive treatment of diabetes on the development and progression of long-term complications in insulin-dependent diabetes mellitus. *New England Journal of Medicine* **329** 977–986.

ENGELHARDT, H. T. (1996). *The Foundations of Bioethics.* Oxford University Press, New York.

FISHER, R. A. (1935). *The Design of Experiments.* Oliver and Boyd, Edinburgh.

FRANCIS, T. (1955). An evaluation of the 1954 poliomyelitis vaccine trials – summary report. *American Journal of Public Health* **45** 1–63.

FREEDMAN, D., PISANI, R., AND PURVES, R. (1998). *Statistics.* W. W. Norton, New York.

FRIEDMAN, L. M., FURBERG, C. D., AND DEMETS, D. L. (2010). *Fundamentals of Clinical Trials.* Springer-Verlag, New York.

HALSEY, N. A., SOMMER, A., HENDERSON, D. A., AND BLACK, R. E. (1997). Ethics and international research. *British Journal of Medicine* **315** 965–966.

KELSEY, J. L., WHITTEMORE, A. S., EVANS, A. S., AND THOMPSON, W. (1996). *Methods in Observational Epidemiology.* Oxford University Press, New York.

LEWIS, E. J., HUNSICKER, L. G., LAN, S., ROHDE, R. D., LACHIN, J. M., and the Lupus Nephritis Collaborative Study Group. (1992). A controlled trial of plasmapheresis therapy in severe lupus nephritis. *New England Journal of Medicine* **326** 1373–1379.

LIND, J. (1753). *A Treatise of the Scurvy.* Sands, Murray, and Cochran, Edinburgh.

LURIE, P. AND WOLFE, S. M. (1997). Unethical trials of interventions to reduce perinatal transmission of the human immunodeficiency virus in developing countries. *New England Journal of Medicine* **337** 853–856.

MATTHEWS, J. N. S. (2006). *An Introduction to Randomized Controlled Clinical Trials.* Chapman and Hall/CRC, Boca Raton, FL.

PALMER, C. R. AND ROSENBERGER, W. F. (1999). Ethics and practice: alternative designs for randomized phase III clinical trials. *Controlled Clinical Trials* **20** 172–186.

PIANTADOSI, S. (2013). *Clinical Trials.* John Wiley & Sons, Inc., New York.

POCOCK, S. J. (1984). *Clinical Trials: A Practical Approach.* John Wiley & Sons, Inc., New York.

ROSENBAUM, P. R. (2002). *Observational Studies.* Springer-Verlag, New York.

SELVIN, S. (2004). *Statistical Analysis of Epidemiologic Data.* Oxford University Press, New York.

Surgeon General. (1982). *The Health Consequences of Smoking: Cancer.* U. S. Department of Health and Human Services, Public Health Service, Rockville, MD.

TYGSTRUP, N., LACHIN, J. M., AND JUHL, E. (1982). *The Randomized Clinical Trial and Therapeutic Decisions.* Marcel Dekker, New York.

2

Issues in the Design of Clinical Trials

2.1 INTRODUCTION

Whereas laboratory science is expected to be performed in a carefully controlled and monitored environment, clinical trials are experiments that are conducted in the workplace of medicine: physician's offices, clinics, or hospitals, as opposed to laboratories. Many clinical trials are *multicenter*, that is, they are performed by a group of participating clinics or care units and, hence, are conducted by a large network of nurses, research coordinators, and physicians. Large amounts of data on study subjects are recorded and computerized. To complicate matters even further, study subjects are human beings, who are often asked to self-administer study treatments at home. So, clinical trials are a complex, collaborative effort involving physicians, nurses, computer scientists, data managers, and statisticians. And guiding everyone in this collaborative effort is the *study protocol*, a document that describes the aims, procedures, and official policies of the scientific endeavor. The importance of the protocol cannot be understated: a study participant who violates the protocol may bias the study and make any conclusions invalid. Consequently, protocol adherence is carefully monitored both internally and externally. In this chapter, we will discuss design issues in clinical trials that every protocol should address.

2.2 STUDY OUTCOMES

The ultimate basis for any scientific investigation is the statement of its objectives. For a clinical trial, the specific aims should be stated so as to define the target population, the time course of observation, and, perhaps most important to the statistician, the outcome measures. Such polemics are easily stated but unfortunately are difficult to

Randomization in Clinical Trials: Theory and Practice, Second Edition.
William F. Rosenberger and John M. Lachin.
© 2016 John Wiley & Sons, Inc. Published 2016 by John Wiley & Sons, Inc.

implement. This stage is crucial, however, because all other design features stem from the statement of objectives, including the statistical analysis plan.

In general, short-term, fixed-duration clinical trials tend to be focused on direct estimation of a treatment effect in terms of the difference of means, rates, or proportions, for example, between the group assigned to the experimental innovation (the treated group) and the control group. Longer-term variable follow-up trials are often focused on the time to some event, such as death, or some measure of disease progression. Such trials are called *survival trials*, a generic term that is used to describe time-to-event outcomes, where the event need not be death, such as time to disease progression, or to remission or even to healing in some cases.

In virtually any disease, there are defined stages of worsening (or improvement) of severity, which are used in clinical research, if not in clinical practice, to describe the stages of disease progression. To the extent possible, the objectives of the clinical trial, and hence its outcome measures, should be defined in terms of clinically relevant indices of disease progression or *clinical effectiveness*.

The statistician's responsibility is to help the investigators to frame the statement of objectives in such a manner that a clinically relevant primary outcome measure and a testable statistical hypothesis are specified, from which the primary statistical analysis is also specified. This primary outcome analysis will drive the design of the study: its length, the number of subjects to be randomized (see Section 2.6), and the statistical analysis plan. Leading a group of investigators to a single primary hypothesis may be one of the greatest challenges that a statistician ever faces. It should be noted that clinical trials are often large enough to answer many other interesting secondary hypotheses, including the effects on secondary outcome measures, or the effectiveness of treatments within subgroups. However, the design should be impelled by a single primary outcome measure of clinical effectiveness.

It is tempting, but dangerous, to plan a clinical trial to only elicit information on the *biological activity* of a therapy (e.g., the effect of a drug on tumor size in cancer or CD4 levels in AIDS). Such information can be elicited quickly and easily. But biological activity is only a *surrogate outcome* for a meaningful outcome of interest in a clinical trial that reflects clinical effectiveness. Clinical effectiveness unequivocally affects patients in a tangible way; for example, by lengthening life (survival time) or increasing quality of life. These outcomes take much longer to ascertain, but a valid clinical trial should be able to determine the true clinical outcome of patients on a therapy.

Prentice (1989) proposed a set of statistical criteria that should be satisfied for concluding that a treatment would favorably affect a clinical outcome based on demonstration of a treatment effect on a biological surrogate outcome. Based on these and other considerations, Fleming and DeMets (1996) present four models in which a surrogate outcome is inappropriate in determining clinical effectiveness:

Model 1. The disease affects the surrogate and the true clinical outcome, but independently. For example, smoking causes yellow fingers and causes lung cancer and death, but an intervention that reverses yellow fingers (the surrogate outcome) may do nothing to reduce premature deaths due to smoking (clinical effectiveness).

Model 2. The disease affects the true outcome via the surrogate, and the intervention bypasses the surrogate. For example, a drug may indeed improve survival (clinical effectiveness) but not have any impact on CD4 counts in AIDS (surrogate).

Model 3. The disease affects the true outcome via a surrogate, but an intervention targeting the surrogate outcome causes adverse effects with respect to the clinical outcome. The literature is replete on drugs that have an effect on biological activity, but have been shown to have no effect or a deleterious effect on survival or other clinical outcome. For example, encainide and flecainide reduced arrhythmias but, relative to placebo, tripled the death rate (Echt *et al.*, 1991).

Model 4. The disease affects the true outcome via a surrogate, but intervention targeting the clinical outcome has no effect on the surrogate. For example, gamma interferon contributed to a 70 percent reduction, relative to placebo, in infections in children with chronic granulomatous disease, yet had no effect on killing bacteria (International Chronic Granulomatous Disease Cooperative Study Group, 1991).

Fleming and DeMets give two criteria for evaluating the relevance of a surrogate outcome in a clinical trial. First, it must be correlated with the clinical outcome. Second, it must fully capture the net effect of the treatment on the clinical outcome. The second criterion is often difficult to determine. Validating a surrogate requires a comprehensive understanding of causal path of disease process and the intervention's intended and unintended effect. Therefore, measures of biological activity should be used with caution as outcomes in clinical trials.

Nevertheless, some definitive outcomes in chronic diseases, such as death, may occur so far in the future that the clinical outcome is not logistically ascertainable. As an example, consider the study of captopril, an angiotensin converting enzyme (ACE) inhibitor, in progressive diabetic nephropathy or kidney disease, one of the complications of diabetes mellitus. The earliest stage of nephropathy is the leakage of tiny amounts of albumin into urine. When the leakage reaches the level that can be detected using an ordinary "dip stick" in urine, about 300 mg/24 hours, the subject has developed overt proteinuria at which point nephropathy is well established. The process will ultimately lead to total destruction of all of the functioning glomeruli that are the biological filters in the kidney and the patient enters renal failure. Life can then be sustained by either dialysis or a renal transplant. Animal studies showed that captopril might reduce the rate of progression to renal failure among patients with proteinuria. This process, however, could take many years, and a clinical trial designed to demonstrate an effect on the incidence of renal failure was considered unfeasible.

In clinical practice, the concentration of creatinine in serum (mg/dL) is universally employed as a simple measure of renal function and to monitor the decline in renal function over time. When the glomerular filtration rate (GFR) falls below the normal range, the level of serum creatinine begins to rise. Although it might be tempting to perform a fixed-duration trial to compare the mean rise in serum creatinine from baseline between the treatment groups, such an analysis would deal only with group means, rather than individual response to therapy. A more appropriate design would be to employ an outcome that is the time to a specific "event" of clinical relevance in individual patients. Such a design would provide a better description of clinical progression in the population than would a simple comparison of means. Further, since each patient is followed to the time of an outcome that represents clinical progression in that individual, consequently the outcome of the trial represents the treatment effect on the incidence of a clinically relevant outcome.

Earlier studies had shown that the inverse creatinine declined linearly over time in patients with established nephropathy (proteinuria). Thus, the study was designed to detect a treatment effect on the time to doubling of the baseline serum creatinine or the time to a 50 percent reduction in the GFR. While no studies were available to show that the Prentice or Fleming–DeMets criteria were satisfied, virtually all physicians would agree that a treatment effect on this outcome is highly meaningful. The trial demonstrated a 48 percent reduction ($p < 0.007$) in the incidence (hazard) of renal progression using this outcome (Lewis *et al.*, 1993). Despite the smaller number of events, a 50 percent reduction was observed in the risk of death or renal transplant ($p < 0.006$).

2.3   SOURCES OF BIAS

The objective of any clinical trial is to provide an unbiased comparison of the differences between treatments. As we shall see in the next chapter, the randomization of subjects between the treatment groups is the paramount statistical element that allows one to claim that a study is unbiased. However, randomization alone does not provide an unbiased study. As Lachin (2011) points out, randomization is necessary, but alone is not sufficient. Two other requirements are: (i) the outcome assessments should be obtained in a like and unbiased manner for all patients; and (ii) data that are missing, if any, from randomized patients do not bias the comparison of the treatment groups. In this section, we describe sources of bias in clinical trials. Some of these biases can be mitigated by randomization, as we shall see in later chapters.

By definition, a clinical trial entails the treatment of patients (or healthy individuals) under scientific conditions. In order for the results to have an impact on the practice of medicine, the treatment procedures employed must be precisely described. For the results to be scientifically rigorous, all aspects of the treatment and clinical management of patients should be standardized as much as possible. These considerations are especially important in a multicenter trial in which it is important that each clinical team ideally should treat and manage patients in an identical manner. This is also important statistically. Virtually everything that happens to a patient after randomization into the trial is a potential outcome measure, especially clinical events that reflect progression of a patient's disease or events that reflect an adverse effect of treatment. For these reasons, to the extent possible, the trial should define and standardize all aspects related to the administration of the study treatments, the ascertainment of clinical events, and clinical management.

All clinical trials should employ a standard system of outcome evaluations in all patients randomized. The objective is to ensure that all subjects are evaluated in an unbiased and precise manner regardless of treatment assignment and the response to treatment. This is most readily achieved by employing a uniform schedule of outcome assessments for all patients with a single central unit for the evaluation of the outcome evaluations in all patients.

2.3.1 Selection and ascertainment bias

Selection bias refers to the bias that can be introduced by selecting particular patients to receive a particular treatment. In a randomized clinical trial, this can occur if the physician knows the randomization sequence, or can guess the treatment being assigned, either consciously or subconsciously. Selection bias has been discussed in the literature since the 1950s, and historical papers (e.g., Blackwell and Hodges, 1957) modeled the impact of a physician's attempt to guess the next allocation and then select a particular patient for a particular treatment.

Berger (2005) distinguishes between *allocation concealment* and *masking*. He defines allocation concealment as the negation of the ability to observe upcoming allocations, so that investigators are not aware of the treatment assignment until after it is assigned. Masking is more complex, as it requires that the treatment assignment not be revealed until the conclusion of the trial. While investigators cannot always be masked in trials, allocation concealment is always possible. Allocation concealment is essential in preventing selection bias, but selection bias can occur even when allocation concealment is present, as investigators can still attempt to guess the treatment to be assigned. Certain randomization procedures, as we shall see, reduce the predictability of treatment assignments and mitigate selection bias further.

To the extent possible, all trials should be *double-masked*, meaning that neither the patient nor the physician is aware of the treatment randomly assigned to the patient. Under no circumstance should the masking be broken, unless there is a serious adverse event that requires knowledge of the assigned treatment. Although it is clear that the patient should not know the treatment assignment, it is often questioned why the physician or caregiver should not be informed. If the physician or recruiting person knows or can guess what treatment will be assigned next, he or she could bias the study by selecting a patient more likely to benefit from the treatment. The physician may also treat patients differently according to which treatment the patient is taking. In these ways, subtle biases can influence the results. These biases can be mitigated by double-masking.

Double-masking may not be possible in some clinical trials, for example, clinical trials of surgical procedures where the surgeon must know which procedure to perform. In unmasked trials, *ascertainment bias* can occur, where the outcome assessment is biased by knowledge of the treatment assignment. If a trial is unmasked to the investigator, it is preferable that outcome evaluations be masked to treatment assignment to the extent possible, where an independent evaluator is unaware of the treatment assignment. For example, in a clinical trial of laser therapy for glaucoma, the outcome might be a deterioration in the visual field. The visual field tests, however, could then be forwarded to a central reading facility where the readers are masked to the treatment assignments of individual eyes. Another example of a clinical trial in which double-masking was impossible is the Diabetes Control and Complications trial, where patients were randomly assigned to receive either conventional or intensive blood glucose control management. However, investigators who evaluated principal outcome measures were masked to treatment assignments. While masking of treatment may not be possible, the evaluation of outcomes can almost always be masked. This is one reason that many trials employ a central laboratory or reading center. Another reason is that, at least theoretically, it is easier to control the accuracy and precision of measurements (i.e., quality control) of a central laboratory than those of multiple laboratories.

2.3.2   Statistical analysis philosophy

There are two prevailing philosophies in the analysis of clinical trials, especially in studies of pharmaceuticals (drugs). On one side is the pharmacologist who wishes to assess the pharmacologic efficacy of the regimen. In this sense, an *efficacy analysis* is performed using the subset of patients who are able to tolerate the drug, are adequately compliant, and to whom the agent is effectively administered. The basic strategy is to examine the experience of the patients entered into the trial and then to select the subset of these patients that meet the desired efficacy criteria for inclusion into the analysis. On the other side is the clinician or regulatory scientist who wishes to assess the overall clinical effectiveness, meaning the outcomes of all patients for whom the treatment is initially prescribed, irrespective of potential side effects or incomplete administration. Although compliance is an important determinant of ultimate effectiveness, the therapeutic question is to assess the effectiveness of the treatment in a population of ordinary subjects with variable degrees of compliance. Such an analysis is called *intention-to-treat*, because the outcome is compared between two samples that are initially assigned to receive different treatments, regardless of the level of tolerance or compliance. Such an analysis attempts to assess the long-term effects in the population of an initial treatment decision to adopt one regimen versus another. In order to conduct an intention to treat analysis, therefore, all subjects randomized into the study must be evaluated as scheduled during follow-up, regardless of the extent of compliance with the treatment protocol or the occurrence of adverse effects.

Following Lachin (2000) is easy to see how bias might enter the study under an efficacy analysis. If one starts a study with 100 patients who are randomized equally between two treatment groups, but at the end of the study, outcome assessments are obtained in only 60 of these, then those 60 patients may not be unbiased. This is because the observations missing for the 40 patients may not be missing completely at random (MCAR), meaning that the presence or absence of an observation occurs *independently of any observed or unobserved data*. For example, a patient on placebo may choose to be noncompliant because he or she feels the treatment is not effective. Likewise, a patient who begins to feel better on an experimental therapy may opt to discontinue medication during the course of the study. In these cases, missingness depends on the outcome of interest, and the remaining subset analysis ignores important information on effectiveness of the treatment in missing patients. Consequently, the only incontrovertibly unbiased study is one in which all randomized patients are evaluated and included in the analysis, and this is the essence of the intent-to-treat philosophy. It should be very clear that final outcome ascertainment should be the investigator's goal for each individual patient enrolled in the study, regardless of their level of active participation in the trial. Thus, the essence of the intention-to-treat design is to ensure that every patient randomized is followed and evaluated as scheduled until either death or the end of the trial.

Thus, the core tenet of the intent-to-treat principle is an intent-to-treat design in which all subjects randomized are followed as specified under the original protocol, regardless of compliance with the treatment regimens, or adverse effects, or whatever; the only exceptions being death, a clinical proscription against the follow-up procedure, or withdrawal of patient consent.

2.3.3   Losses to follow-up and noncompliance

Various authors have used the terms "losses to follow-up," "dropouts," and "noncompliance" interchangeably. In this book, we use the term *lost to follow-up* to describe patients who do not continue follow-up visits for primary outcome assessments. Such patients may have moved away or may no longer be willing to participate in the study for various reasons. Provided that the reason lost to follow-up is not related to the outcome of the study, the data on these patients are MCAR and should not bias the study. Every effort should be made to ascertain the reason these patients dropped out of the study and if it is at all treatment-related. Adjustments to the number of patients randomized are usually built into the study to accommodate a small number of losses to follow-up, as we will see in Section 2.6.4. However, such adjustment compensates for the loss of information, not the potential bias that can be introduced by missing data. By *noncompliance*, we refer to less than maximally effective treatment in a patient who continues follow-up. These patients should be included in an intention-to-treat analysis, to avoid bias.

Even when placebo controls are used to implement double-masking, some drugs, for example, have a known adverse effect profile, such as where the drug is known to induce mild hepatotoxicity, or gastrointestinal disturbances, and so on. In such cases, substantial biases may be introduced if subjects who experience the adverse effects are terminated from further follow-up. The resulting missing observations are not MCAR, and it is not possible to argue that the resulting observed measures are unbiased.

A recent report from the National Research Council (2010) reviews the importance of avoiding missing data as much as possible, and many of the model-based approaches available to attempt to adjust for the presence of missing data.

2.3.4   Covariates

When comparing two treatment groups, it is often desired to ensure that baseline characteristics of patients are similar in each group. A comparison of baseline characteristics is usually presented in Table 1 of medical articles presenting the results of a clinical trial. Balance over baseline characteristics is presented as evidence that the "randomization worked as intended." Cosmetic balance on baseline covariates is often deemed to be critical in making a convincing case that the treatment was effective. In Chapters 7 and 9, we will describe randomization techniques designed to force certain known covariates to be balanced.

We shall also see in later chapters, however, that balancing on known baseline covariates is not always the statistically most efficient design. As Cornfield pointed out in Section 1.3, one of the important properties of randomization is that it tends to balance on both known and *unknown* covariates. This property implies that, in a perfectly randomized clinical trial, a significant imbalance is the result of making a type I error. Hence, tests of significance comparing baseline characteristics between treatment groups are not really tests that randomization worked, but that the trial was properly randomized without any bias (Senn, 1994). This led Berger (2005) to conclude that testing baseline covariate imbalances is in reality testing the presence or absence of selection bias. Senn (1994) also notes that investigators rarely take into account the multiple testing problem in testing multiple covariates and that the absence of a significant p-value does not imply balance.

Efron (1971) developed a model to describe the impact of ignoring an important covariate as a main effect in a regression model of the treatment effect. He called this impact *accidental bias*. Certain randomization procedures can mitigate accidental bias, as we shall see in Chapter 4.

In addition to assessing or describing the comparability of treatment groups with respect to known covariates, measured covariates are also used to assess the association of covariate values with the outcome of the study, and also to assess the treatment group effect as a function of covariate values, such as separately between men and women; we call these *treatment-by-covariate interactions*. Randomization has no impact on whether treatment-by-covariate interactions will exist since these are true characteristics of the phenomena under study, not the result of chance.

The Diabetes Prevention Program (Diabetes Prevention Program Research Group, 2002) showed that treatment with the drug metformin versus placebo provides a 31 percent reduction ($p < 0.001$) in the risk of developing type 2 diabetes in individuals with impaired glucose intolerance. Among the important analyses was an assessment of this treatment effect among subjects stratified into subgroups defined by the baseline level of body mass index (BMI) in kg/cm^2. Among those with BMI < 30, metformin provided only a 3 percent risk reduction versus placebo, whereas the risk reduction was 16 percent among those with $30 \leq$ BMI < 35, and 53 percent among those with BMI $\geq$ 35. The heterogeneity of treatment effect among these BMI subgroups was significant at $p < 0.05$. Thus, balancing treatment groups on the level of BMI would not alone lead to correct conclusions: that the drug was effective only in a certain subgroup of patients and not in another subgroup. If such a subgroup analysis is known to be important in advance, studies can be powered accordingly to detect a treatment by covariate interaction,, but in practice, studies often do not have enough power to detect such interactions, and such subgroups may not be known in advance. Consequently, it is important to remember that, while randomization tends to induce independence between the treatment effect and unobserved covariates by eliminating accidental bias, it does not eliminate interactions.

2.4   EXPERIMENTAL DESIGN

The essence of a clinical trial is the comparison of the effects of the experimental treatment to those of a control treatment. For example, in a therapeutic trial to compare a stated dose of a drug versus a placebo, if the other aspects of the trial are rigorous with regard to controlling bias, then a sharp comparison can be made to discern the clinical effects of the drug. Although there may be multiple factors that contribute to the effectiveness of an experimental treatment, trials are not usually designed to elucidate those precise factors, but only to determine if the treatment is effective. In most trials, therefore, the experimental design is quite simple. Usually only two or more groups are employed to compare two or more treatments. The treatment effect can, for example, be estimated from a simple one-way group comparison.

While *factorial designs* (*cf.* Cochran and Cox, 1957) are frequently employed in other sciences, they are rarely used in clinical trials. The exception is the consideration of combination therapies, where a two-way factorial design may be employed. An analysis of variance strategy is then employed, where treatment A and treatment B are main effects and there is an interaction term for treatments A and B. The test of interaction assesses whether the combination has effects above and beyond each treatment individually. However, such studies are often designed with the assumption that there is no interaction and are then underpowered for a test of interaction. If an interaction is later observed, then the effects of each level of A depend on the level of B and vice versa. In this case, the study also is underpowered for the detection of nested effects of factor A within each level of B and vice versa. In addition, if a binary or time-to-event outcome analysis is employed, even when there is no interaction but a marginal treatment effect on one factor is present, the power for the assessment of the treatment effect on the other factor will be reduced. Rarely are higher order factorial designs employed.

Short-term clinical trials of chronic, but stable, conditions will sometimes employ a *crossover design* (*cf.* Jones and Kenward, 2014), whereby each patient will receive one of two treatments initially, and then, after a *washout period* of withdrawal from the treatment, the patient receives the other treatment. For the analysis of this design, the results of treatment A are compared to those of treatment B in aggregate over all patients. Thus, the marginal statistic for treatment A actually consists of the effect of treatment A during the first period of administration plus the residual effect of treatment A during the second period of administration (after having received treatment B during the first period). So one must assume that there are no residual carryover effects from the first period of administration to the second period of administration. It also must be possible to withdraw the experimental therapy from some patients in the second phase, so obviously surgical trials and many therapeutic trials would be excluded from consideration. The advantage of crossover trials is that they employ fewer patients, since each patient serves as his or her own control. However, the condition of no carryover effect is difficult to satisfy, and consequently, they are rarely used in long-term clinical trials, which are designed to assess the effects of treatment on the clinical course of a disease.

A wholly different concept is *cluster randomization*, in which clusters are randomized rather than individual subjects. In some cases, randomizing individuals may be infeasible or particularly costly. It may also be that the treatment effect across clusters is considered more important scientifically than the treatment effect across individuals and it is desired to measure a cluster effect. However, the analysis is complicated since underlying differences among clusters is confounded with the treatment effect. There are several books devoted to cluster randomization (e.g., Donner and Klar, 2010), and such trials require special considerations that will not be covered in this book.

Often in cluster randomized trials, a *stepped wedge* design is employed, which is similar to a crossover trial with K time points, but once assigned the experimental treatment, a cluster can never revert to the control. In time period 1, all $K - 1$ clusters are assigned to control. In time period 2, one cluster is randomized to treatment and the other clusters to control. In time period 3, a second cluster is randomized to treatment. This continues until all $K - 1$ clusters are assigned to the experimental treatment in time period K. A stepped wedge design is most appropriate when crossing over from experimental therapy to control is either ethically or logistically infeasible. It also allows the detection of trends over time and duration of therapy. While the stepped wedge design can be applied to individuals as well as clusters, it has most often been applied in cluster randomized trials.

Special trial designs are sometimes employed for ethical considerations. In *personal preference randomized designs*, patients are randomized to either a "choice arm" or a "random arm." Those randomized to the random arm are randomized in a standard two-arm parallel structure. Those randomized to the choice arm have the opportunity to choose the treatment to be assigned. Personal preference designs induce complicating factors that make analysis more difficult. The treatment effect is confounded by the selection effect, the expected treatment difference between patients who would choose different treatments if allowed to do so, and a preference effect, the expected treatment difference between participants who do or do not receive their preferred treatment. These considerations are described in Walter *et al.* (2011).

In a *randomized withdrawal design*, all eligible patients are treated with the open-label experimental therapy, and then the subgroup of "responders" (those who reach a predefined threshold of response) are then randomized in a standard two-arm parallel trial. Such designs eliminate from the trial patients who are unlikely to respond to treatment and also minimize the amount of time subjects are exposed to a placebo control (*cf.* Temple, 1996).

For crossover designs, stepped wedge designs, personal preference randomized designs, and randomized withdrawal designs, the distinguishing issue is when randomization takes place; but when randomization is done, it is typically performed in a standard two-arm or multi-arm parallel fashion. Hence, the actual process of randomization requires no special considerations in addition to the important considerations that any randomized clinical trial must take into account.

In conclusion, the most widely employed design for clinical trials is the simple two-group comparison design. Therefore, it is in this context that this book will address the randomization of subjects to receive one of the study treatments. In practice, randomization in multi-arm trials can use existing randomization techniques for two-arm trials, with some modifications. These modifications will be discussed as well.

2.5   RECRUITMENT AND FOLLOW-UP

In a typical clinical trial, patients are identified for screening and consideration for entry into the trial over a period of time, often years. The interval is called the *recruitment period* during which patients are screened and, if found eligible, are then randomized to one of the study treatments. Eligibility requirements are agreed upon by the investigators before the trial begins and are recorded in the study protocol. Eligibility requirements, established according to both ethical and clinical considerations, are designed to ensure that a homogeneous strain of patients is recruited for the trial. If eligibility criteria are changed or relaxed at any point in the trial, shifts in patient characteristics may occur, which can cause problems in the statistical analysis and interpretation of the study results. Likewise, if protocol violations occur where an investigator randomizes an ineligible patient, serious biases may result. The process of recruitment of patients over an interval of time results in what is termed *staggered entry*, because all patients do not enter the trial simultaneously at a given point in calendar time. The clinical trial then systematically collects observations over time according to a follow-up schedule. The follow-up schedule specifies the duration of treatment of each patient and the precise times during follow-up at which specific procedures are performed and measurements obtained.

The two common plans for the design of clinical trials are *fixed* and *variable* follow-up duration. Each provides for a period of recruitment in calendar time of length R, where time 0 is the calendar date at which the first patient enters the trial and time R is the subsequent date on which the last patient is randomized. The trial is then continued until calendar time $T, T > R$, which provides for the follow-up of the last patient entered. The difference between the time a subject enters a trial and follow-up is completed is the *study time* of that subject. In a fixed follow-up duration trial, each patient is followed for the same prescribed period of time, regardless of when that patient was entered into the study. Thus, each patient's study time is the same, and the study cannot be concluded until each patient completes that study time. Fixed-duration trials are often employed to assess short-term objectives, such as when each patient is treated with a drug for a 2-month interval. Many long-term studies employ variable follow-up duration, where patients are followed until a common closing date, regardless of when they were randomized. Thus, a patient's study time depends on the date the patient entered the trial. For example, in a 5-year trial $(T = 5)$ with a 2-year recruitment period $(R = 2)$, the first patient entered would be followed for all $T = 5$ years, whereas the last patient entered would be followed for $T - R = 3$ years. Assuming that recruitment follows a uniform distribution, average duration of follow-up would be 4 years with a standard deviation of 0.57 years.

Uniform recruitment assumes that the distribution function G of patient entry times is linear over $[0, R]$. It is not unusual, though, for G to be convex or concave. Convexity implies that recruitment is initially faster than expected under uniform entry and then declines; concavity implies that recruitment is initially slower than expected and then increases. One possible model is the truncated exponential distribution, where we assume that patient entry times $Z_1, \ldots, Z_n$ are independent and identically distributed with density function

$$g(z) = \frac{\gamma e^{-\gamma z}}{1 - e^{-\gamma R}}, 0 \le z \le R, \gamma \ne 0. \tag{2.1}$$

If $\gamma > 0$, G is convex, and if $\gamma < 0$, G is concave. Under model (2.1), expected duration of follow-up is given by

$$T - \frac{1 - e^{-\gamma R} - \gamma R e^{-\gamma R}}{\gamma(1 - e^{-\gamma R})}, \tag{2.2}$$

with a standard deviation of

$$\frac{(1 + e^{-2\gamma R} - (2 + \gamma^2 R^2)e^{-\gamma R})^{1/2}}{\gamma(1 - e^{-\gamma R})}. \tag{2.3}$$

In our example with $R = 2$ and $T = 5$, if $\gamma = 1$, average duration of follow-up is 4.31 years with a standard deviation of 0.53 years. If $\gamma = -1$, average duration of follow-up is 3.69 years with a standard deviation of 0.53 years (note the symmetry).

With either fixed-duration or variable follow-up, the duration of follow-up should be based on the period of time needed to assess the trial's objectives. The frequency of follow-up visits is usually based on conventional clinical practice for the treatment and follow-up of the condition under study. However, in other instances, the frequency of assessment may be based on other considerations, such as the frequency required to chart the incidence of an event or a change in a characteristic over time or to safeguard patient safety by toxicity screens. During these visits, ascertainment of the study outcome, medical assessments to determine safety of the therapy, determination of compliance with the study's treatment regimen, and any other medical procedures dictated by the standard of care for the condition will be performed. Insofar as possible, clinical trials should mimic usual clinical practice.

2.6   DETERMINING THE NUMBER OF RANDOMIZED SUBJECTS

In the planning stages of a randomized clinical trial, it is necessary to determine the numbers of subjects to be randomized. While the exact final number that contributes to any analysis will be unknown, due to losses to follow-up and staggered entry, it is still desirable to determine a target sample size based on some model. This *sample size estimate* will then allow estimates of the total cost of the trial, the number of clinics required, and target recruitment numbers, and so on. Typically, the number of subjects is computed to provide a fixed level of power under a specified alternative hypothesis (see, e.g., Lachin, 1981 and Donner, 1984). The alternative hypothesis usually represents a minimal, clinically meaningful treatment effect. Power (1− probability of a type II error) is an important consideration for several reasons. Low power can cause a truly beneficial therapy to be rejected. However, too much power may make results statistically significant that are not clinically significant. Standard regulatory criteria for clinical trials often lead to specifying the probability of type I error (α) to be 0.05 and power to be 0.80–0.90. However, such specifications belie the consideration of the relative cost of a type I or type II error in a particular study. There are examples of studies where investigators determined that a type II error was so much more costly than a type I error, that α was fixed far from 0.05 (see, e.g., Samuel-Cahn and Wax, 1986).

Tests of the treatment effect in clinical trials are typically two-sided, for two principal reasons: first, it is usually relevant if the placebo or standard therapy is more efficacious than the experimental therapy; and, second, even if only a one-sided hypothesis is really of interest, a two-sided test requires a more stringent 0.025-level test, which gives added protection from a type I error.

2.6.1   Development of the main formula

Under this framework of power considerations, it is necessary to assume a *population model*. Let n be the total number of subjects randomized in the trial, and let n_i be the number randomized to treatment group i. For two treatments $i = A, B$, say, $n = n_A + n_B$. We assume here that the allocation proportions are known in advance, that is, that $Q = n_A/n$ and $1 - Q = n_B/n$ are predetermined. This will not be the case under adaptive randomization, as we will see later in the book. Under a population model, it is assumed that responses $Y_{ij}, j = 1, \ldots, n_i$, are independently and identically distributed according to some known distribution $G(y_{ij}|\theta_i)$, where θ_i is possibly a vector-valued parameter associated with the ith population. For example, if θ_i is a single parameter representing an outcome associated with treatment, a standard hypothesis test would be $H_0 : \theta_A = \theta_B$ versus $H_1 : \theta_A \neq \theta_B$. Let S_n be a statistic to test a hypothesis regarding the equality of one or more members of θ. Based on the distribution of the measurements, $Y_{ij} \sim G(y_{ij}|\theta_i)$, it is usually easy to derive the distribution of a statistic S_n under H_0 and H_1. The central limit theorem will usually lead to a normal distribution under H_0 and H_1, such as the following:

$$H_0 : S_n \sim N(\mu_0, \Sigma_0^2), \ H_1 : S_n \sim N(\mu_1, \ \Sigma_1^2), \tag{2.4}$$

where μ_0 and μ_1 are functions of θ_i and Σ_0 and Σ_1 are functions of n and θ_i. This provides a large sample test of H_0 of the form $T_n = (S_n - \mu_0)/\Sigma_0$, which is asymptotically distributed as standard normal under H_0. With this test, H_0 is rejected at level α if $|T_n| \geq z_{\alpha/2}$ (two-sided) or $T_n \geq z_\alpha$ (one-sided), where z_α is the standard normal deviate; that is, if Z is a standard normal variate, $\Pr(Z \geq z_\alpha) = \alpha$. The power of the test Z is given by $1 - \beta$. Figure 2.1 shows the relationship between α and β for the standard hypothesis testing problem.

The basic relationship used to derive n, based on values of β and α under a specified alternative hypothesis, can be derived as follows. Under the distributional assumption

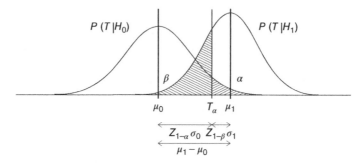

Fig. 2.1 *Distribution of a test statistic T under the null and alternative hypotheses, with the rejection region of size α and a type II error rate of size β.*

in (2.4), for a one-sided alternative ($\mu_1 > \mu_0$),

$$1 - \beta = \Pr\left(T_n > z_\alpha | H_1\right)$$
$$= \Pr\left(\frac{S_n - \mu_0}{\Sigma_0} > z_\alpha | H_1\right)$$
$$= \Pr\left(S_n > \mu_0 + z_\alpha \Sigma_0 | H_1\right)$$
$$= \Pr\left(\frac{S_n - \mu_1}{\Sigma_1} > \frac{\mu_0 + z_\alpha \Sigma_0 - \mu_1}{\Sigma_1} \left| \frac{S_n - \mu_1}{\Sigma_1} \sim N(0, 1)\right.\right)$$
$$= \Pr\left(Z > \frac{\mu_0 - \mu_1}{\Sigma_1} + z_\alpha \frac{\Sigma_0}{\Sigma_1}\right), \tag{2.5}$$

where Z is a standard normal variate. This implies that

$$-z_\beta = \frac{\mu_0 - \mu_1}{\Sigma_1} + z_\alpha \frac{\Sigma_0}{\Sigma_1}.$$

Simple algebra then leads to the equation

$$\mu_1 - \mu_0 = z_\alpha \Sigma_0 + z_\beta \Sigma_1. \tag{2.6}$$

For a two-sided test, (2.6) is given by

$$|\mu_1 - \mu_0| \cong z_{\alpha/2} \Sigma_0 + z_\beta \Sigma_1. \tag{2.7}$$

If the variance does not depend on the mean parameters, as in normal variates, then $\Sigma_0 = \Sigma_1$ in formula (2.5) for power. In the Poisson and binomial cases, this will not be true, but the result then relies on the central limit theorem, and Σ_0/Σ_1 behaves as $0(1)$. However, in finite samples, we can still estimate this term and use it as an adjustment. Since the common value of the parameters is not specified under the null hypothesis, it is often the convention to substitute the mean of the parameter values specified in the alternative hypothesis.

For more than two groups, formulas can be adjusted accordingly, and a Bonferroni correction is standardly used for more than two hypothesis tests. For example, in the Medical Therapy of Prostatic Symptoms (MTOPS) trial (McConnell *et al.*, 2003), patients were randomized to one of four groups: placebo, finasteride, doxazosin, or finasteride and doxazosin. Each of the active therapy groups was to be compared to the placebo group, for a total of three hypothesis tests. Hence, for a two-sided test, a Bonferroni adjustment led to the term $z_{\alpha/6}$ in equation (2.7).

2.6.2 Example

Consider a comparison of two means. Here we assume $Y_{ij} \sim N(v_i, \sigma_i^2), i = A, B, j = 1, \ldots, n_i, n_A = Qn$. Assume $\sigma_A^2 = \sigma_B^2 = \sigma^2$, and σ^2 is known. Then $\theta_i = (v_i, \sigma^2)$. We wish to test $H_0 : v_A = v_B$ versus $H_A : v_A \neq v_B$, so that $\mu_0 = 0$ and $\mu_1 = v_A - v_B$, using the statistic $S_n = \bar{y}_A - \bar{y}_B$, where $\bar{y}_i = \sum_{j=1}^{n_i} y_{ij}/n_i$. Then $\Sigma_0^2 = \Sigma_1^2 = [Q(1-Q)]^{-1} \sigma^2/n$, and equation (2.7) becomes

$$|\mu_1| = \frac{(z_{\alpha/2} + z_\beta)\sigma}{\sqrt{Q(1-Q)n}} \qquad (2.8)$$

or

$$n = \frac{1}{Q(1-Q)} \left(\frac{(z_{\alpha/2} + z_\beta)\sigma}{\mu_1} \right)^2. \qquad (2.9)$$

Similar expressions can be derived for differences of binomial probabilities or Poisson rates (Problems 2.4 and 2.5).

This example presents the statistician with two dilemmas: σ^2 is assumed to be known, which is always an unreasonable assumption, and we can only compute the required number of subjects under a specified alternative. For both problems, the statistician must rely on the physician who is familiar with the natural history of the disease under no treatment (in the case of a placebo-controlled trial) or under the standard therapy (in a trial comparing a new and existing therapy). The physician must provide an estimate of σ^2 using information from previous studies or from his or her experience. The particular alternative specified should represent the minimal clinically significant difference for which the physician would declare the experimental therapy to be successful or worthwhile in practice.

2.6.3 Survival trials

In a survival trial where the study objective is to compare a time-to-event outcome between two treatment groups, equation (2.7) can be used to derive the number of randomized subjects under a parametric failure-time model. Lachin and Foulkes (1986) assume an exponential model for testing the equality of two hazard functions,

λ_A and λ_B. Since the log-rank test for equality of hazard (survival) functions over time is known to be the asymptotically most efficient test under a proportional hazards model, and since the exponential distribution is the special case of proportional constant hazards, the exponential model leads to estimates of the sample size and power function of the log-rank test. If the actual hazard in the control group fluctuates over time, calculations based on the exponential model will still be adequate provided that the expected number of events in the control group under the model agrees with the actual expected value. If not, then alternate methods, such as those of Lakatos (1988), are preferred.

For the test $H_0 : \lambda_A = \lambda_B$ versus $H_A : \lambda_A \neq \lambda_B$, $S_n = \hat{\lambda}_A - \hat{\lambda}_B$, the maximum likelihood estimator of $\mu_1 = \lambda_A - \lambda_B$. Let $\bar{\lambda} = Q\lambda_A + (1 - Q)\lambda_B$. Under H_0, the asymptotic distribution is $N(0, \Sigma_0^2)$, and under H_1, the asymptotic distribution is $N(\mu_1, \Sigma_1^2)$, where $\Sigma_0^2 = \phi(\bar{\lambda})/nQ(1 - Q)$ and $\Sigma_1^2 = ((1 - Q)\phi(\lambda_A) + Q\phi(\lambda_B))/nQ(1 - Q)$. The function ϕ will depend on the patient entry distribution and losses to follow-up. Substituting into equation (2.7), we obtain

$$n = \frac{\left\{ z_{\alpha/2}\left(\phi(\bar{\lambda})\right)^{1/2} + z_\beta((1 - Q)\phi(\lambda_A) + Q\phi(\lambda_B))^{1/2} \right\}^2}{\mu_1^2 Q(1 - Q)}. \tag{2.10}$$

One would use z_α in place of $z_{\alpha/2}$ for a one-sided test. Note that this equation requires specification of λ_A and λ_B, which can be estimated if one knows something about the cumulative incidence functions on treatments A and B. The hazard function can be computed from the incidence rate over T years, ρ, by the formula $\lambda = -\log(1 - \rho)/T$. If B is placebo, we can compute an estimate of λ_B based on knowledge of the cumulative incidence of the disease on no therapy and then compute the hazard rate λ_A on the experimental treatment assuming a specific reduction in risk. For instance, if λ_B is estimated to be 0.05, then if we wish to detect a 33 percent reduction in risk, $\lambda_A = (0.67)(0.05) = 0.0335$.

Under uniform recruitment over $[0, R]$ and variable follow-up over $(T - R, T)$ with no losses to follow-up, Lachin (1981) shows that

$$\phi(\lambda) = \lambda^2 / \Pr(\text{event}|R, T, \lambda)$$
$$= \lambda^2 \left(1 - \frac{e^{-\lambda(T-R)} - e^{-\lambda T}}{\lambda R} \right)^{-1}. \tag{2.11}$$

In this and subsequent expressions, $\Pr(\text{event}|R, T, \lambda)$ is the probability of the event in a cohort recruited over R years and followed over T years with hazard rate λ. Thus, the power of the test and the required sample size depend on these probabilities. When substituted into (2.10), this yields the required sample size needed to provide a given number of events in each group. These required numbers of events are virtually identical for other study plans specified by the values of R and T.

Lachin and Foulkes (1986) derive the expression with an adjustment for losses to follow-up under the assumption that losses to follow-up are random in each group and time to loss to follow-up is independent of the survival or event times. They consider the special case where losses are exponentially distributed with hazard rates η_A and η_B for groups A and B, respectively. Assuming uniform recruitment, equation (2.10) becomes

$$n = \frac{(\text{term1} + \text{term2})^2}{\mu_1^2 Q(1 - Q)},$$

where

$$\text{term1} = z_{\alpha/2}\left((1 - Q)\phi(\bar{\lambda}, \eta_A) + Q\phi(\bar{\lambda}, \eta_B)\right)^{1/2},$$

$$\text{term2} = z_{\beta}\left((1 - Q)\phi(\lambda_A, \eta_A) + Q\phi(\lambda_B, \eta_B)\right)^{1/2},$$

and

$$\phi(\lambda, \eta) = \lambda^2 \left(\frac{\lambda}{\eta + \lambda} \left(1 - \frac{e^{-(T-R)(\eta+\lambda)} - e^{-T(\eta+\lambda)}}{R(\eta + \lambda)}\right)\right)^{-1}. \tag{2.12}$$

Note that, when $\eta = 0$, (2.12) reduces to (2.11).

Now suppose that patient entry times are distributed as truncated exponential, as in (2.1). Lachin and Foulkes (1986) show that this entry distribution yields

$$\phi(\lambda, \gamma) = \lambda^2 \left(1 + \frac{\gamma e^{-\lambda T}(1 - e^{(\lambda-\gamma)R})}{(1 - e^{-\gamma R})(\lambda - \gamma)}\right)^{-1}. \tag{2.13}$$

If we employ the exponential entry distribution in conjunction with exponentially distributed losses to follow-up, we obtain

$$\phi(\lambda, \eta, \gamma) = \lambda^2 \left(\frac{\lambda}{\eta + \lambda} + \frac{\lambda \gamma e^{-(\lambda+\eta)T}(1 - e^{(\lambda+\eta-\gamma)R})}{(1 - e^{-\gamma R})(\lambda + \eta)(\lambda + \eta - \gamma)}\right)^{-1}. \tag{2.14}$$

Equation (2.14) reduces to (2.13) if $\eta = 0$.

Of course, survival may not follow an exponential distribution; other failure distributions include the Weibull and log-normal, for instance. Similar formulas could have been derived under other failure time distributions. However, it is unlikely that the investigators will have some knowledge of the form of the survival distribution *a priori*, and hence, these computations can only be considered an approximation based on our current knowledge.

Finally, it should be noted that the aforementioned computations for sample size n provides the number of subjects that are expected to provide the desired level of power under the specified alternative hypothesis. In the survival setting, such trials are called *maximum-duration designs*. An alternate approach is a *maximum-information design*, in which subjects are recruited and followed until the amount of statistical information is obtained that guarantees the desired level of power. In the survival setting, such designs are called *event-driven*, since the amount of information required is a function of the numbers of subjects that reach the event. See Lachin (2005) for more details.

2.6.4 Adjustment for noncompliance

As we discussed in Section 2.3.2, compliance is not an issue if one is interested in the clinical effectiveness of a therapy, that is, the effectiveness in the general population that includes subjects who may not comply with their prescribed regimen. If a study is of pharmacologic efficacy, such noncompliant subjects are often terminated from further follow-up, or their data excluded from analysis. Of course, this admits the strong potential for bias. However, if the objective of the study is true effectiveness, then under an intent-to-treat design, all such noncompliant subjects would continue to be followed and their outcome data used in all analyses. In this case, noncompliance can severely compromise power relative to an intent-to-treat effectiveness trial in which all patients are fully compliant. Similar considerations apply when inappropriate patients are entered into the trial, such as those who may be misdiagnosed, who would not benefit from treatment even if the patient were fully compliant.

To see this, let $Q = 0.5$ and let the proportion of noncompliant or inappropriate patients in the experimental group be ω and let the hazard rate be λ_A. These noncompliant patients are then assumed to have the same hazard rate as the control group. Also, assume that noncompliant patients in the control group (e.g., placebo) will continue to have the same hazard rate (e.g., there is no placebo effect), given by λ_B. Then μ_1^* is the expected treatment effect under noncompliance, given by $\mu_1^* = (1 - \omega)\lambda_A + \omega\lambda_B - \lambda_B = (1 - \omega)\mu_1$. Assuming that the asymptotic variances are similar in the two groups, we obtain the following adjustment from equation (2.10):

$$n^* = \frac{n}{(1 - \omega)^2}.$$

Note that this is quite a substantial adjustment. A noncompliance rate of 10 percent will require randomizing 23 percent more patients, Note again that this assumes that all n subjects are followed and scheduled.

In most clinical trials, the hazards are specified and the corresponding risk reduction are specified in terms of the overall rates in the general population, recognizing that some fraction will be noncompliant. In that case, noncompliance is already allowed for in the estimates. In other cases, however, it may be desirable to specify the hazards and the risk reduction, assuming 100 percent compliance. Then an adjustment such as the aforementioned could be used to factor for noncompliance, assuming complete follow-up. However, this is a severe adjustment because it assumes that a noncompliant subject has the same hazard as a control subject, regardless of how long the subject may have actually complied before becoming noncompliant, and assumes that biologically any exposure to the experimental treatment short of complete 100 percent compliance has no effect. These are implausible assumptions, and Thus, the aforementioned is a worst-case adjustment.

2.6.5 Additional considerations

While sample size calculations are a required element of proposals and protocols for randomized clinical trials, it is important to note the deficiencies of this approach.

First, the formulas derived depend on unknown parameters whose values must be guessed. While the objective is to describe the sample size required to provide a desired level of power to detect a difference between groups that is clinically relevant, the actual computation requires specification of other unknown parameters. For example, in the case of two normal means, we must rely on a specification of the variance; in the case of a survival trial, we must specify the incidence of death or progression in the control group. In the latter case, for a placebo-controlled clinical trial, a substantial *placebo effect* might be present, and so these guesses are likely to be wrong. For example, suppose the hazard rate in the placebo is expected to be 0.05, but is really 0.04 due to a positive placebo effect, and we wished to detect $\mu_1 = 0.02$. With the placebo effect, we would really need to detect $\mu_1 = 0.01$, which would require a fourfold increase in sample size to detect with the same power. If our guesses are too far off, we could severely overestimate power and wind up with negative results for an effective experimental therapy. It is important to emphasize that such guesses should be conservative and calculations should be conducted over a range of values. Although it might be tempting to be as economical as possible in determining the number of subjects for an expensive clinical trial, this approach is foolhardy.

Second, these computations rely on a population model whereby individuals are assumed to be sampled at random from respective populations. Later, we introduce another approach to conducting a test of significance that is based on a randomization model that considers the probabilities of treatment assignment and their covariances, if any. Randomization models have advantages over population models, and if a randomization model is to be adopted for the final analysis of a trial, then sample size calculations based on a population may not be correct and should be viewed only as an approximation. But one can consider every aspect of sample size computation an approximation, because one must guess the underlying distribution and the underlying variability. In later chapters, we will discuss the distinction between population and randomization models for the randomized clinical trial and how this might affect power.

2.7 PROBLEMS

2.1 Write a protocol for a hypothetical clinical trial. The trial will consist of a new therapy for a known disease versus a placebo. Search the medical literature for information on similar studies on the disease. Such studies should provide information on estimated incidence rates, loss to follow-up rates, information on primary outcome measures, and follow-up schedule. The protocol should include eligibility criteria, primary and secondary outcomes, study time and considerations of fixed versus variable duration, statistical analysis philosophy, and numbers of subjects randomized.

2.2 Derive equations (2.2) and (2.3) from equation (2.1).

2.3 Prove equation (2.7) from first principles. Give an intuitive explanation as to why the z_β term is unaffected for a two-sided test.

2.4 Consider independent and identically distributed observations from a Poisson distribution with rate parameter λ. The maximum likelihood estimator of λ is $\hat{\lambda} = T/n$, where T is the total number of events in n units, such as T epileptic seizures in n patient-years of exposure. Now consider two groups with parameters λ_A and λ_B with sample sizes Qn and $(1 - Q)n$, respectively, $0 < Q < 1$.

a. Derive the basic expression relating sample size and power for the test of difference between the two groups, using the formula in (2.7).
b. Consider a study to compare a drug versus placebo in the treatment of epileptics. What parameter will have to be estimated from prior knowledge?

2.5 Consider the case of two simple proportions with expectations π_A and π_B. We wish to plan a study to assess healing with an investigational drug (A) and placebo (B).

a. Derive the basic expression relating sample size and power for the two-sided test of difference of probabilities of healing between the two groups. Assume $n_A = Qn$ and $n_B = (1 - Q)n$.
b. Prior studies suggest that the control healing rate is on the order of 20 percent. Investigators believe that a minimal, clinically meaningful increase in healing on the experimental therapy is 5 percent. For 80 percent power, compute the number of patients needed, assuming equal allocation $(Q = 0.5)$.
c. From part (b), investigate the changes in n that occur with changing the allocation proportions.

2.6 For a clinical trial comparing two normal means, as presented in Section 2.6.2, suppose the standard deviation on treatment A is σ_A and the standard deviation on treatment B is σ_B, where $\sigma_A \neq \sigma_B$. Show that, for fixed n, the value of Q that maximizes power is given by $Q^* = \sigma_A/(\sigma_A + \sigma_B)$. (Allocating according to the ratio of the standard deviations is called *Neyman allocation*. This result implies that equal allocation does not maximize power when the two treatments have different standard deviations.)

2.7 The MTOPS trial (McConnell *et al.*, 2003) was a variable follow-up trial with $R = 2$ and $T = 6$, designed with four treatment groups: placebo (group I), finasteride (group II), doxazosin (group III), and combination of finasteride and doxazosin (group IV). The primary outcome is three comparisons: I versus II, I versus III, and I versus IV with respect to a time-to-progression outcome. Compute the number of randomized subjects needed for these comparisons for a 50 percent reduction in risk, when the incidence rate over 5 years is assumed to be 25 percent. Make the following assumptions: $\alpha = 0.05$ (two-sided), 80 percent power, exponential incidence, and uniform recruitment. Build in a 10 percent exponential loss to follow-up rate over the 5 years.

2.8 REFERENCES

BERGER, V. W. (2005). *Selection Bias and Covariate Imbalance in Randomised Clinical Trials.* John Wiley, Chichester.

BLACKWELL, D. and HODGES, J. L. (1957). Design for the control of selection bias. *Annals of Mathematical Statistics* **28** 449–460.

COCHRAN, W. G. and COX, G. M. (1957). *Experimental Designs*. John Wiley & Sons, Inc., New York.

Diabetes Prevention Program Research Group. (2002). Reduction in the incidence of type 2 diabetes with lifestyle intervention or metformin. *New England Journal of Medicine* **346** 393–403.

DONNER, A. (1984). Approaches to sample size estimation in the design of clinical trials – a review. *Statistics in Medicine* **3** 199–214.

DONNER, A. and KLAR, N. (2010). *Design and Analysis of Cluster Randomization Trials in Health Research*. John Wiley & Sons, Inc., New York.

ECHT, D. S., LIEBSON, P. R., MITCHELL, L. B., PETERS, R. W., OBIAS-MANNO, D., BARKER, A. H., and for the Cardiac Arrhythmia Suppression Trial. (1991). Mortality and morbidity in patients receiving encainide, flecainide, or placebo. *New England Journal of Medicine* **324** 781–788.

EFRON, B. (1971). Forcing a sequential experiment to be balanced. *Biometrika* **58** 403–417.

FLEMING, T. R. and DEMETS, D. L. (1996). Surrogate end points in clinical trials: are we being misled? *Annals of Internal Medicine* **125** 605–613.

International Chronic Granulomatous Disease Cooperative Study Group. (1991). A controlled trial of interferon gamma to prevent infection in chronic granulomatous disease. *New England Journal of Medicine* **324** 509–516.

JONES, B. and KENWARD, M. G. (2014). *Design and Analysis of Crossover Trials*. Chapman and Hall/CRC, Boca Raton, FL.

LACHIN, J. M. (1981). Introduction to sample size determination and power analysis for clinical trials. *Controlled Clinical Trials* **2** 93–113.

LACHIN, J. M. (2000). Statistical considerations in the intent-to-treat principle. *Controlled Clinical Trials* **21** 167–189.

LACHIN, J. M. (2005). Maximum information designs. *Clinical Trials* **2** 453–464.

LACHIN, J. M. and FOULKES, M. A. (1986). Evaluation of sample size and power for analyses of survival with allowance for nonuniform patient entry, losses to follow-up, noncompliance, and stratification. *Biometrics* **42** 507–519.

LAKATOS, E. (1988). Sample sizes based on the log-rank statistic in complex clinical trials. *Biometrics* **44** 229–241.

LEWIS, E. J., HUNSICKER, L. G., BAIN, R. P., ROHDE, R. D., and the Collaborative Study Group. (1993). The Effect of Angiotensin-Converting-Enzyme Inhibition on Diabetic Nephropathy. *New England Journal of Medicine* **329** 1456–1462.

MCCONNELL, J. D., ROEHRBORN, C. G., BAUTISTA, O. M., ANDRIOLE, G. L., DIXON, C. M., KUSEK, J. W., LEPOR, H., MCVARY, K. T., NYBERG, L. M., CLARKE, H. S., CRAWFORD, E. D., DIOKNO, A., FOLEY, J. P., FOSTER, H. E., JACOBS, S. C., KAPLAN, S. A., KREDER, K. J., LIEBER, M. M., LUCIA, M. S., MILLER, G. J., MENON, M., MILAM, D. F., RAMSDELL, J. W., SCHENKMAN, N. S., SLAWIN, K. M., SMITH, J. A., and for the Medical Therapy of Prostatic Symptoms (MTOPS) Research Group. (2003). The long-term effect of doxazosin, finasteride, and combination therapy on the clinical progression of benign prostatic hyperplasia. *New England Journal of Medicine* **349** 2387–2398.

National Research Council. (2010). *The Prevention and Treatment of Missing Data in Clinical Trials*. The National Academies Press, Washington, DC.

PRENTICE, R. L. (1989). Surrogate endpoints in clinical trials: definition and operational criteria. *Statistics in Medicine* **8** 431–440.

SAMUEL-CAHN, E. and WAX, Y. (1986). A sequential test for comparing two infection rates in a randomized clinical trial, and incorporation of data accumulated after stopping. *Biometrics* **42** 99–108.

SENN, S. (1994). Testing for baseline balance in clinical trials. *Statistics in Medicine* **13** 1715–1726.

TEMPLE, R. (1996). Problems in interpreting active control equivalence trials. *Accountability in Research* **4** 267–275.

WALTER, S. D., TURNER, R. M., MACASKILL, P., MCCAFFERY, K. J., and IRWIG, L. (2011). Optimal allocation of participants for the estimation of selection, preference, and treatment effects in the two-stage randomised trial design. *Statistic in Medicine* **31** 1307–1322.

3

Randomization for Balancing Treatment Assignments

3.1 INTRODUCTION

Thus far, we have talked quite loosely about randomization as the toss of a fair coin, but simple coin tossing is rarely employed in a clinical trial. One can distinguish four classes of randomization procedures: *complete randomization, restricted randomization, covariate-adaptive randomization*, and *response-adaptive randomization*.

Let $T_1, \ldots, T_n$ be a sequence of random treatment assignments, where $T_i = 1$ if the patient is assigned to treatment A and $T_i = 0$ if the patient is assigned to treatment B, $i = 1, \ldots, n$. *Complete randomization* is simple coin tossing, in which case, $T_1, \ldots, T_n$ are independent and identically distributed Bernoulli random variables with $p = \Pr(T_i = 1) = 1/2, i = 1, \ldots, n$. In *restricted randomization procedures*,[1] $T_1, \ldots, T_n$ are dependent, with variance–covariance matrix given by $\Sigma_T \neq (1/4)I$. Restricted randomization is employed when it is desired to have equal numbers of patients assigned to each treatment group (i.e., balancing the treatment assignments), and this will be the topic of this chapter. The dependence structure characterizes the procedure, as all valid restricted randomization procedures will have the property that $E(T_i) = 1/2, i = 1, \ldots, n$ (and hence $\mathrm{Var}(T_i) = 1/4$), a property referred to by Kuznetsova and Tymofyeyev (2011) as "preserving the allocation ratio at every allocation." However, the conditional expectation will define the different procedures.

---

[1] Yates, in his discussion of a paper by Anscombe (1948), was perhaps the first to use the term *restriction* in the context of randomization, but this was in the context of agricultural experiments. He referred to certain valuable, but complicated variants of Latin-squares designs, with additional restrictions that would limit the usual number of sequences.

Randomization in Clinical Trials: Theory and Practice, Second Edition.
William F. Rosenberger and John M. Lachin.
© 2016 John Wiley & Sons, Inc. Published 2016 by John Wiley & Sons, Inc.

One can also use restricted randomization for fixed unbalanced allocation, as we shall see.

Covariate-adaptive randomization is used when it is desired to ensure balance between treatment arms with respect to certain known covariates. Treatment assignments will depend on the covariate values of patients. Finally, *response-adaptive randomization* is used when ethical considerations make it undesirable to have equal numbers of patients assigned to each treatment arm. In response-adaptive randomization, the treatment assignments depend upon previous patient responses to treatment. Covariate-adaptive and response-adaptive randomization will be treated in later chapters. The four types of randomization procedures are progressively more complicated, from a statistical point of view, due to the increased complexity of the dependence structure. A generalization of response-adaptive randomization, covariate-adjusted response-adaptive (CARA) randomization, incorporates a subject's covariate profile into response-adaptive randomization. CARA randomization will be considered as a special case of response-adaptive randomization. It is dealt with more extensively in Chapter 7 of Hu and Rosenberger (2006).

In this chapter, we discuss complete randomization and restricted randomization procedures for balancing treatment assignments. Friedman, Furberg, and DeMets (1981, p. 41) present two main arguments for equal allocation to treatment groups. The first is that power is maximized when allocation is equal. The second is that equal allocation is consistent with the concept of equipoise that should exist at the beginning of the trial. Many clinical trialists disagree, in principle, with these arguments, and we will explore alternative arguments in later chapters on response-adaptive randomization. However, it should be noted that most clinical trials today do employ restricted randomization procedures to achieve balance, and these arguments have become rooted, to some extent, in the culture of clinical trials.

We now present the principal randomization tools to achieve balance among the treatment groups. A thorough probabilistic analysis of these randomization procedures is required. In particular, the conditional expectation of assignment, given all previous assignments, will define the procedure. The unconditional variance–covariance structure of the treatment assignments is used to develop the theoretical susceptibility to bias of each procedure (Chapters 4 and 5) as well as to determine the distribution of randomization-based inferential tests (Chapter 6).

3.2 COMPLETE RANDOMIZATION

When treatment assignments are independent Bernoulli random variables with success probability $1/2$, we have complete randomization. Complete randomization has some very nice properties, in that certain types of bias are minimized. For example, there can be no selection bias with complete randomization, since it is equally likely to guess the next treatment assignment correctly or incorrectly. However, there is a disadvantage to complete randomization that makes it unattractive in practice: there is a nonnegligible probability of some imbalances between treatments and a small probability of severe imbalances. In fact, the theory of probabilities of

Table 3.1   *Percentiles of the distribution of* $|D_n|$ *for complete randomization.*

n	0.33	0.25	0.10	0.05	0.025
50	6.9	8.1	11.6	13.9	15.9
100	9.7	11.5	16.5	19.6	22.4
200	13.8	16.3	23.3	27.7	31.7
400	19.5	23.0	32.9	39.2	44.8
800	27.6	32.5	46.5	55.4	63.4

large deviations should serve as a warning when using Bernoulli sequences for randomization in small to moderate samples.

Let $N_A(i) = \sum_{j=1}^{i} T_j, i = 1, \ldots, n$ so that $N_A(i)$ is the number of patients randomized to treatment A after i patients have been randomized. Let $N_B(i) = i - N_A(i)$. Then by the central limit theorem for a binomial random variable, $N_A(n)$ is asymptotically normal with mean $n/2$ and asymptotic variance $n/4$. Letting $D_n = N_A(n) - N_B(n) = 2N_A(n) - n$, we see that D_n is asymptotically normal with mean 0 and variance n. We can use $|D_n|$ as one measure to describe the degree of imbalance between treatment groups. For $r > 0$,

$$\Pr(|D_n| > r) \cong 2\{1 - \Phi(r/\sqrt{n})\}, \tag{3.1}$$

where Φ is the standard normal distribution function. One can use this formula to determine the degree to which complete randomization is subject to imbalances of size r, for a large sample trial with n patients (see Problem 3.3). Table 3.1 gives percentiles of the distribution of $|D_n|$ for various values of n. For example, when $n = 50$, there is a 5 percent chance of an imbalance of ± 13.9 or worse. This corresponds to an excess of 6.95 beyond the expected 25 in either group, or an imbalance of 36.1 percent versus 63.9 percent. When $n = 400$, there is a 5 percent chance of an imbalance of ± 39.2, corresponding to a degree of imbalance of 45.1 percent versus 54.9 percent. To many, an imbalance of this degree would be of no concern.

While some imbalances will likely occur, apart from cosmetic concerns, the important question is whether these imbalances compromise the statistical properties of the study. Regardless of the final sample sizes, balanced or not, the resulting estimate of the treatment group difference will still be unbiased. While an imbalance will decrease the precision of the estimator, this effect will be slight for moderate imbalances.

Likewise, an imbalance will decrease the power of statistical test, but again, the effect is slight for moderate imbalances. For example, consider the example in Section 2.6.2, the comparison of two normal means with equal variances. Power can be computed using equation (2.8). We draw the power curves across values of Q for $n = 50, 100, 200$ in Figure 3.1, where $\sigma = 1$ and $|\mu_1| = 0.5$. For large n, the curve flattens at the top, indicating that there is little loss of power for Q between 0.30 and 0.70. As shown in Table 3.1, the probability of a large imbalance following complete randomization is very small. This has led some investigators to recommend

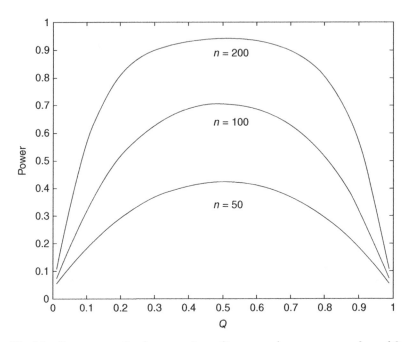

Fig. 3.1   *Power curves for the comparison of two normal means across values of Q.*

complete randomization as a matter of course for trials of moderate to large size (Schultz and Grimes, 2002 recommend $n > 200$). However, the probability of a large imbalance is even smaller with restricted randomization designs, and many restricted randomization procedures protect against biases as well.

3.3   FORCED BALANCE PROCEDURES

Forced balance procedures are restricted randomization procedures that lead to exactly $n/2$ patients assigned to A and $n/2$ to B. While the investigator may have control over the total number of subjects to be randomized in small animal studies or phase I trials, it is rarely possible in a phase III clinical trial. Therefore, forced balance procedures are seldom used as standalone procedures; however, they are routinely applied in blocks.

3.3.1   Random allocation rule

Let F_n be the set of treatment assignments through the first n stages of the randomization process, that is, $F_n = \{T_1, \dots, T_n\}$. (For a formal mathematical treatment, F_n is a *sigma-algebra* and F_0 is the trivial sigma-algebra.) Note that $\Pr(T_n = 1) = E(T_n)$.

Then the random allocation rule is defined by the following allocation probabilities:

$$E(T_j|\mathcal{F}_{j-1}) = \frac{\frac{n}{2} - N_A(j-1)}{n - (j-1)}, j = 2, \ldots, n, \tag{3.2}$$

and $E(T_1) = 1/2$. We call $E(T_j|\mathcal{F}_{j-1})$ the *allocation rule* for the restricted randomization procedure. Note that the allocation rule is itself a random variable and is computed at the realizations of $T_1, \ldots, T_j$. For example, if patient 50 is ready to be randomized in a clinical trial of $n = 100$ patients, and thus far, the allocation has been 28 to A and 21 to B. Then patient 50 will receive treatment A with probability $22/51$.

One can immediately see some problems with this rule. First, once $n/2$ patients have been assigned to one treatment, all further treatment allocations are deterministic, and hence absolutely predictable, and selection bias can result. Second, at some stage in the middle of the trial, there could be significant treatment imbalances. If patients entering the trial are heterogeneous with respect to some important covariate related to outcome (e.g., if there is a time trend), then imbalances between treatment groups with respect to that covariate may result. For example, suppose $n_1 < n$ patients have been randomized in the clinical trial. Conditional on n_1, $N_A(n_1)$ follows a hypergeometric distribution with mean $n_1/2$ and variance $n_1(n - n_1)/4(n - 1)$. The exact probability of imbalance of $N_A(n_1)$ to $N_B(n_1)$ can be obtained from Fisher's exact test for the resulting 2×2 table with cells $N_A(n_1), N_B(n_1), n/2 - N_A(n_1)$, and $n/2 - N_B(n_1)$. Asymptotically, the distribution of $N_A(n_1)$ can be approximated by a normal distribution. Let $D_{n_1,n} = N_A(n_1) - N_B(n_1)$. For $r > 0$,

$$\Pr\left(|D_{n_1,n}| > r\right) \cong 2\left\{1 - \Phi\left(r\sqrt{(n-1)/n_1(n-n_1)}\right)\right\}. \tag{3.3}$$

Clearly, this function is maximized when $n_1/n = 0.5$.

One can think of the random allocation rule in terms of an urn model. Suppose an urn contains $n/2$ balls of type A and $n/2$ balls of type B. Each time a patient is ready to be randomized, a ball is drawn and *not* replaced, and the corresponding treatment is assigned. This continues until the urn is depleted. One can easily see that this leads to the allocation rule in (3.2).

With the urn formulation, we can see that random allocation rule produces $\binom{n}{n/2}$ equally likely permutations of $n/2$ A's and $n/2$ B's. The unconditional probability of treatment assignment, given by $E(T_j)$, can be found by thinking of the jth element of the $\binom{n}{n/2}$ permutation sequences. Of these sequences, half are A's and half are B's. Let $I(T_{jl} = 1)$ be 1 if the jth patient in the lth sequence is assigned to treatment A, 0 otherwise. Let p_l be the probability of the lth sequence (in the random allocation rule, $p_l = 1/\binom{n}{n/2}$ for all $l = 1, \ldots, L$, where $L = \binom{n}{n/2}$). Then

$$E(T_j) = \sum_{l=1}^{L} I(T_{jl} = 1)p_l = \left(\binom{n}{n/2}\Big/2\right)\Big/\binom{n}{n/2} = \frac{1}{2}. \tag{3.4}$$

Since $T_j^2 = T_j$, it follows that $\mathrm{Var}(T_j) = E(T_j) - \{E(T_j)\}^2 = 1/4$, as for complete randomization. The differences between complete randomization and restricted randomization are completely specified by $\mathrm{cov}(T_i, T_j)$, which is 0 for complete randomization. For the random allocation rule, we can compute, for $i < j$,

$$
\begin{aligned}
E(T_i T_j) &= \Pr(T_i = 1, T_j = 1) \\
&= \Pr(T_j = 1 | T_i = 1) \cdot E(T_i) \\
&= \frac{\frac{n}{2} - 1}{n - 1} \cdot \frac{1}{2} \\
&= \frac{n - 2}{4(n - 1)}.
\end{aligned}
$$

Consequently,

$$
\mathrm{cov}(T_i, T_j) = \frac{n - 2}{4(n - 1)} - \frac{1}{4} = \frac{-1}{4(n - 1)}. \tag{3.5}
$$

3.3.2   Truncated binomial design

An alternate way of assigning exactly $n/2$ patients to each treatment is to randomly allocate each according to the toss of a coin until one treatment has been assigned $n/2$ times; all subsequent patients will receive the opposite treatment. Blackwell and Hodges (1957) refer to this as the *truncated binomial design*. We can describe the design by the rule

$$
\begin{aligned}
E(T_j | \mathcal{F}_{j-1}) &= \frac{1}{2}, && \text{if} \quad \max\{N_A(j - 1), N_B(j - 1)\} < \frac{n}{2}, \\
&= 0, && \text{if} \quad N_A(j - 1) = \frac{n}{2}, \\
&= 1, && \text{if} \quad N_B(j - 1) = \frac{n}{2},
\end{aligned}
$$

for n even. As in the random allocation rule, the tail of the randomization sequence will be entirely predictable and there can be serious imbalances during the course of the trial.

We can quantify the number of subjects in the tail whose treatment assignment is deterministic using discrete distribution theory (Blackwell and Hodges, 1957). Let X be the random number of subjects in the tail. Then $X = x$ can occur if the $n/2$th A assignment or the $n/2$th B assignment occurs for patient $n - x$, for n even. These events have probability

$$
\frac{1}{2^{n-x}} \binom{n - x - 1}{\frac{n}{2} - 1},
$$

according to the negative binomial distribution. Therefore,

$$\Pr(X = x) = \frac{1}{2^{n-x-1}} \binom{n-x-1}{\frac{n}{2}-1}, \quad x = 1, \ldots, \frac{n}{2}, \tag{3.6}$$

a truncated negative binomial distribution. From this distribution, one can derive

$$E(X) = \frac{n}{2^n} \binom{n}{n/2} \tag{3.7}$$

and

$$E(X^2) = n - E(X) \tag{3.8}$$

(Problem 3.4). The results (3.7) and (3.8) will be needed later in Chapter 5.

We now compute the unconditional mean, variance, and covariance of the treatment assignment indicators. This can be accomplished by conditioning on the random variable $\tau = \min\{i : \max(N_A(i), N_B(i)) = n/2\}$. It is clear that $\Pr(N_A(\tau) = n/2) = \Pr(N_B(\tau) = n/2) = 1/2$. Conditional on $N_A(\tau) = n/2$, we compute the probability of assignment to A as follows:

$$\begin{aligned} E(T_j) &= EE(T_j|\tau) \\ &= E\Pr(T_j = 1|\tau) \\ &= \frac{1}{2} \cdot \Pr(\tau \geq j) + 0 \cdot \Pr(\tau < j). \end{aligned} \tag{3.9}$$

Similarly, conditional on $N_B(\tau) = n/2$, we compute

$$\begin{aligned} E(T_j) &= E\Pr(T_j = 1|\tau) \\ &= \frac{1}{2}\Pr(\tau \geq j) + 1 \cdot \Pr(\tau < j). \end{aligned} \tag{3.10}$$

Since (3.9) and (3.10) each have probability $1/2$, we compute the unconditional expectation as

$$\begin{aligned} E(T_j) &= \frac{1}{4} \cdot \Pr(\tau \geq j) + \frac{1}{2} \cdot \Pr(\tau < j) + \frac{1}{4} \cdot \Pr(\tau \geq j) \\ &= \frac{1}{2}. \end{aligned}$$

Therefore, $\mathrm{Var}(T_j) = 1/4$.

It remains to compute $\text{cov}(T_i, T_j)$. Again, we break the problem into two pieces by conditioning first on $N_A(\tau) = n/2$. For $i < j$,

$$E(T_i T_j) = EE(T_i T_j | \tau)$$
$$= E \Pr(T_i = 1, T_j = 1 | \tau)$$
$$= \frac{1}{4} \cdot \Pr(\tau \geq j) + 0 \cdot \Pr(\tau < j).$$

Similarly, conditioning on $N_B(\tau) = n/2$, we obtain

$$E(T_i T_j) = E \Pr(T_i = 1, T_j = 1 | \tau)$$
$$= \frac{1}{4} \cdot \Pr(\tau \geq j) + \frac{1}{2} \cdot \Pr(i \leq \tau < j) + 1 \cdot \Pr(\tau < i).$$

Unconditioning, we see that

$$E(T_i T_j) = \frac{1}{4} \cdot \Pr(\tau \geq j) + \frac{1}{4} \cdot \Pr(i \leq \tau < j) + \frac{1}{2} \cdot \Pr(\tau < i)$$
$$= \frac{1}{4} + \frac{1}{4} \cdot \Pr(\tau < i).$$

Consequently,

$$\text{cov}(T_i, T_j) = \frac{1}{4} \cdot \Pr(\tau < i). \tag{3.11}$$

The distribution of τ comes from (3.6), as $\Pr(X = x) = \Pr(\tau = n - x)$, and hence

$$\text{cov}(T_i, T_j) = \frac{1}{4} \sum_{k=n-i+1}^{n/2} \frac{1}{2^{n-k-1}} \binom{n-k-1}{\frac{n}{2}-1}. \tag{3.12}$$

Note that $\text{cov}(T_i, T_j) = 0$ if $i < (n+2)/2$.

While the truncated binomial design and the random allocation rule both yield $\binom{n}{n/2}$ permutations of As and Bs, the sequences will not be equiprobable for the truncated binomial design. This is demonstrated in Table 3.2 for $n = 4$.

3.3.3 Hadamard randomization

A *Hadamard matrix* is a matrix of 1's and 0's in which two different rows have the same entries in half the columns and opposite entries in the other half. Construction of Hadamard matrices and its existence in various sizes are topics for algebraists specializing in the theory of automorphism groups. For our purposes, a Hadamard matrix exists and can be written down explicitly if n is divisible by 4 for $n \leq 424$ (Hedayat, Sloane, and Stufken, 1999). (e.g., the function "Hadamard" in the R software package (Lumley, 2012) returns a Hadamard matrix for a specific value of n.) Bailey and

Table 3.2 *Six permutation sequences for n = 4 with probabilities under truncated binomial randomization (each sequence has probability 1/6 under random allocation).*

Sequence	Probability
AABB	$\dfrac{1}{4}$
ABAB	$\dfrac{1}{8}$
ABBA	$\dfrac{1}{8}$
BABA	$\dfrac{1}{8}$
BAAB	$\dfrac{1}{8}$
BBAA	$\dfrac{1}{4}$

Nelson (2003, p. 556) describe one such Hadamard matrix with one row removed that they deem useful for creating a randomization procedure with $n = 12$:

$$
H = \begin{bmatrix}
0 & 1 & 0 & 0 & 1 & 0 & 1 & 0 & 0 & 1 & 1 & 1 \\
0 & 1 & 1 & 1 & 0 & 0 & 1 & 1 & 0 & 1 & 0 & 0 \\
0 & 0 & 1 & 0 & 1 & 0 & 0 & 1 & 1 & 1 & 1 & 0 \\
0 & 1 & 0 & 1 & 1 & 1 & 0 & 1 & 0 & 0 & 1 & 0 \\
0 & 1 & 1 & 1 & 0 & 0 & 0 & 0 & 1 & 0 & 1 & 1 \\
0 & 1 & 1 & 0 & 1 & 1 & 1 & 0 & 1 & 0 & 0 & 0 \\
0 & 0 & 1 & 1 & 1 & 1 & 0 & 0 & 0 & 1 & 0 & 1 \\
0 & 0 & 0 & 1 & 1 & 0 & 1 & 1 & 1 & 0 & 0 & 1 \\
0 & 0 & 0 & 1 & 0 & 1 & 1 & 0 & 1 & 1 & 1 & 0 \\
0 & 1 & 0 & 0 & 0 & 1 & 0 & 1 & 1 & 1 & 0 & 1 \\
0 & 0 & 1 & 0 & 0 & 1 & 1 & 1 & 0 & 0 & 1 & 1
\end{bmatrix}. \tag{3.13}
$$

Using this matrix, they develop a procedure, which they call *Hadamard randomization*, by creating $2(n - 1)$ equiprobable randomization sequences. They suggest doubling the number of rows with the mirror image of the rows of H (i.e., changing the 0's to 1's and the 1's to 0's in the concatenated rows). This can be thought of as a restriction of the random allocation rule to include only $2(n - 1)$ of the $\binom{n}{n/2}$ sequences. For $n = 12$, this reduces the set of possible sequences from 924 to 22. The advantage of the Hadamard matrix in (3.13) is that it eliminates the possibility of extreme sequences (e.g., *AAAAAABBBBBB*), which are undesirable if the treatment difference is confounded with a time trend. It also eliminates sequences such as *ABABABABABAB*. However, due to the ambiguous structure of the Hadamard matrix, it is not clear how the randomization sequence can be generated using a standard allocation rule $E(T_j | \mathcal{F}_{j-1})$, making many theoretical properties impossible to obtain, such as the variance–covariance structure of the treatment assignments.

3.3.4 Maximal procedure

The random allocation rule produces $\binom{n}{n/2}$ sequences, and the maximum imbalance is $n/2$. Berger, Ivanova, and Knoll (2003) proposed the *maximal procedure* to restrict the number of sequences to those that have a maximum imbalance of b. As with the random allocation rule, each sequence is equiprobable, and the procedure can be thought of as a random walk with a reflecting barrier b steps off the diagonal. The random walk (i, D_i) starts at $(0, 0)$ and ends at $(n, 0)$. Let $\#(k, D_k)$ be the number of paths between (k, D_k) and $(n, 0)$, such that $|D_k| \le b, k = 0, \ldots, n-1$, with $D_0 = 0$. The number of paths required is solved using graph theoretic methods. Then one can define the procedure as

$$E(T_i | \mathcal{F}_{i-1}) = \frac{\#(i, D_i)}{\#(i-1, D_{i-1})}. \tag{3.14}$$

Notice that the numerator of (3.14) is a function of D_i, and therefore cannot be observed, but the symmetry of the random graph can be applied to compute the probabilities. The number of sequences with a given initial sequence is the same as the number of sequences with the initial sequence reversed at the end of the sequence. For example, for $b = 3$ and $n = 12$, there are 792 sequences in the maximal procedure. Then 396 sequences end with A and 396 sequences end with B, so the same number must start with A and B. So, 396 sequences go from $(1, 1)$ to $(n, 0)$ and 396 sequences from $(1, -1)$ to $(n, 0)$. This implies that $E(T_1 | \mathcal{F}_0) = 396/792$. Of these 396 sequences, there are 232 sequences that end with AB, so there are 232 sequences that start with BA, and hence with AB, so there are $464/2 = 232$ sequences that go from $(2, 0)$ to $(n, 0)$. Likewise, there are $396 - 232 = 164$ sequences that start with AA and go from $(2, 2)$ to $(n, 0)$, and 164 sequences that start with BB and go from $(2, -2)$ to $(n, 0)$. Then

$$E(T_2 | \mathcal{F}_1) = \frac{\#(2, 2)}{\#(1, 1)} = \frac{164}{396}, \quad \text{if} \quad D_1 = 1,$$

$$= \frac{\#(2, 0)}{\#(1, -1)} = \frac{232}{396}, \quad \text{if} \quad D_1 = -1.$$

In order to find the number of sequences, an algorithm has been developed by Salama, Ivanova, and Qaqish (2008).

3.4 FORCED BALANCE RANDOMIZATION WITHIN BLOCKS

3.4.1 Permuted block design

Complete randomization, the random allocation rule, and the truncated binomial design can all result in severe imbalances at some point during the trial. This is particularly undesirable if there is a time-heterogeneous covariate that is related to treatment outcome, because imbalances in treatment assignments can then lead to imbalances in those important covariates. To avoid this, *permuted block designs* are often used to ensure balance throughout the course of the clinical trial, by imposing a balance restriction at various stages in the trial. Permuted block designs are reviewed by Zelen (1974).

For the permuted block design, we establish M blocks containing $m = n/M$ patients, where M and n/M are positive integers, and within block i, $m/2$ patients are assigned to treatment A and $m/2$ patients are assigned to treatment B. To ensure balance, a random allocation rule is typically used within each block (although one could also use a truncated binomial design), where the total number of patients is m instead of n. When permuted blocks are used, at M stages during the course of the trial, we achieve balanced allocation. The maximum imbalance at any time during the trial is given by $\max_j |D_j| = m/2$.

One could also use a truncated binomial design to achieve complete balance within each block. In this case, the underlying assignment sequence probabilities, and the covariance matrix of the assignments, would be different from those of a random allocation rule.

In the extreme case, $M = n/2$ and every pair randomized is balanced. However, this procedure requires every even randomization to be deterministic, and hence, selection bias is easy to occur, unless the pairs are first identified and then randomized as a set. For these reasons, block sizes larger than 2 are generally employed, and investigators should be masked to the block size selected.

Typically, n is not known in advance, and the final block may be unfilled. If that is the case, the maximum imbalance that can occur in the trial is $m/2$. If blocks are filled by Hadamard randomization, for blocks of size $m = 12$, the maximum imbalance that can occur is 3. However, the trade-off is that the set of possible sequences reduces from 924 to 22 within each block. We refer to the set of all possible sequences generated by a randomization procedure as a *reference set*. Having a smaller reference set has implications in the potential for selection bias and randomization-based inference, as we shall discuss in later chapters.

If $b = m/2$ in the maximal procedure, it will exhibit the same maximum imbalance as the permuted block design. However, the maximal procedure will produce more possible sequences in the reference set than the corresponding permuted block design (filled using the random allocation rule) when there are no unfilled blocks. To see this, let $N_{MP,b,n}$ be the number of possible sequences in the maximal procedure and $N_{PBD,b,n}$ be the number of possible sequences in the permuted block design when $b = m/2$. Then $N_{PBD,b,n} = \binom{2b}{b}^{n/2b}$. Because all admissible sequences for the permuted block design have maximal imbalance b and perfect balance after n patients, any sequence in the reference set must also be in the set of admissible sequences for the maximal procedure. So, $N_{PBD,b,n} < N_{MP,b,n}$. The ratio $N_{MP,b,n}/N_{PBD,b,n}$ increases in n (Problem 3.6).

3.4.2   Random block design

Some biostatisticians and regulatory agencies advocate varying the block size in order to reduce the chance of selection bias, since, if the block size is known, some patients will be assigned to a treatment with probability 1. When clinical trial manuscripts indicate that block sizes were "varied," or "alternated," the procedure is imprecise. A more formal procedure is the *random block design*, in which block sizes are randomly selected from a discrete uniform distribution.

We can define the procedure rigorously as follows. Let $B_j, j = 1, \ldots, n$ be the block size of the block containing the jth patient, where B_j is a random variable that can take the values $2, 4, 6, \ldots, 2B_{max}$, and B_{max} is one half the largest block size. Each block size selected with probability $1/B_{max}$ before the block is filled. Note that for n patients, each block will be filled with the possible exception of the last; n does not need to be known in advance. Given B_j, the position number R_j is defined as the position of the jth patient in a block of size B_j and takes the values $1, 2, \ldots, B_j$. If $R_n = B_n$, then every block is filled and $D_n = 0$ at the end of the trial. Under this procedure, the allocation rule depends on the random variable pairs (B_j, R_j). Filling each block using the random allocation rule, we have

$$E(T_j | F_{j-1}, B_j, R_j) = \frac{\frac{B_j}{2} - \sum_{l=j+1-R_j}^{j-1} T_l}{B_j - R_j + 1} \tag{3.15}$$

(where the sum is zero if the lower summand exceeds the upper). The parameter of interest in this procedure is B_{max}. As the number of possible random block sizes increases, the number of sequences in the reference set increases, but also the probability of a final imbalance as large as B_{max} can result if the last block is unfilled.

Other variations on the random block design are possible. Blocks of size 2 could be eliminated or block sizes could be selected according to some probability distribution that weights block sizes unequally. As a variant on forced balance designs, if one knows the exact number of patients to be enrolled, possible block sizes and be chosen in advance, and all possible permutations of these block sizes that add to exactly n can be selected as the set of possible block sequences (e.g., Heussen, 2004).

3.5   EFRON'S BIASED COIN DESIGN

Efron (1971) developed the *biased coin design* in order to balance treatment assignments. Let D_n be an increasing function of $N_A(n)$ such that $D_n = 0$ if $N_A(n) = n/2$ (e.g., $D_n = N_A(n) - N_B(n) = 2N_A(n) - n$). He suggests allocating with the following rule. Define a constant $p \in (0.5, 1]$.

$$\begin{aligned} E(T_j | F_{j-1}) &= \frac{1}{2}, &&\text{if} \quad D_{j-1} = 0, \\ &= p, &&\text{if} \quad D_{j-1} < 0, \\ &= 1 - p, &&\text{if} \quad D_{j-1} > 0. \end{aligned}$$

The procedure is denoted as $BCD(p)$. We can use a symmetry argument to determine the unconditional probability of assignment to A, $E(T_j)$. By symmetry (and since

$D_0 = 0$), we have

$$\Pr(D_{j-1} < 0) = \Pr(D_{j-1} > 0),$$

and both equal $1/2$ if $j - 1$ is odd. Hence, for $j - 1$ odd, we have

$$E(T_j) = p \Pr(D_{j-1} < 0) + (1 - p) \Pr(D_{j-1} > 0) = \frac{1}{2}.$$

For $j - 1$ even, we have

$$\Pr(D_{j-1} < 0) + \Pr(D_{j-1} > 0) = 2 \Pr(D_{j-1} < 0) = 1 - \Pr(D_{j-1} = 0),$$

and hence,

$$\begin{aligned} E(T_j) &= \frac{\Pr(D_{j-1} = 0)}{2} + p \Pr(D_{j-1} < 0) + (1 - p) \Pr(D_{j-1} > 0) \\ &= \frac{\Pr(D_{j-1} = 0)}{2} + \frac{[1 - \Pr(D_{j-1} = 0)]}{2} \\ &= \frac{1}{2}. \end{aligned}$$

Using the theory of random walks, we can measure the degree of imbalance, $|D_n| = |2N_A(n) - n|$. We have the following transition probabilities:

$$\Pr(|D_{n+1}| = 1 | |D_n| = 0) = 1,$$

and, for a positive integer j,

$$\Pr(|D_{n+1}| = j - 1 | |D_n| = j) = p,$$
$$\Pr(|D_{n+1}| = j + 1 | |D_n| = j) = 1 - p.$$

This yields the following random walk matrix for $|D_n|$:

$$P = \begin{bmatrix} 0 & 1 & 0 & 0 & 0 & \cdots \\ p & 0 & 1-p & 0 & 0 & \cdots \\ 0 & p & 0 & 1-p & 0 & \cdots \\ 0 & 0 & p & 0 & 1-p & \cdots \\ \vdots & \vdots & \vdots & \vdots & \vdots & \end{bmatrix}.$$

We can solve the steady-state equations, by solving for the left eigenvector of P (cf. Karlin and Taylor (1975, p. 86)). These equations are given by

$$\pi_0 = p\pi_1,$$
$$\pi_1 = \pi_0 + p\pi_2,$$
$$\pi_2 = (1 - p)\pi_1 + p\pi_3,$$
$$\vdots \qquad \vdots$$

and $\pi_0 + \pi_1 + \pi_2 + \ldots = 1$. The solution is

$$\pi_0 = \frac{r-1}{2r},$$

$$\pi_j = \frac{(r+1)(r-1)}{2r^{j+1}}, j \geq 1, \tag{3.16}$$

where $r = p/(1-p)$ (Problem 3.7). Since $|D_n|$ can take only odd or even values as n is odd or even, the Markov chain has period 2, and the π's must be doubled (*cf.* Ross (1983, p. 111)). We can obtain the limiting balancing property as

$$\lim_{m \to \infty} \Pr(|D_{2m}| = 0) = 2\pi_0 = 1 - \frac{1}{r},$$

$$\lim_{m \to \infty} \Pr(|D_{2m+1}| = 1) = 2\pi_1 = 1 - \frac{1}{r^2}.$$

(Note that for odd n, the minimum imbalance is 1.) Obviously, as $p \to 1$, we achieve perfect balance, but such a procedure is deterministic. When $p = 2/3$, we have probability of $1/2$ of achieving perfect balance for even n and probability $3/4$ with odd n for large n.

We can also derive the steady-state variance of D_n using (3.16) (Problem 3.8). It is given by:

$$\frac{4r(r^2 + 1)}{(r^2 - 1)^2}, \quad \text{when number of trials is even}$$

$$\frac{8r^2}{(r^2 - 1)^2} + 1, \quad \text{when number of trials is odd.} \tag{3.17}$$

The difference between odd and even number of trials is due to the differences in the supports of the distributions: in particular, a significant mass is concentrated at 0 when n is even and p is large. The steady-state distribution of the Markov chain induces a constant variance, which differs from complete randomization, where the variance of D_n grows at a rate $O(n)$.

Markaryan and Rosenberger (2010) derive $P(D_n = k), -n \leq k \leq n$ exactly using combinatorial methods; it is also derived by Haines (2013) using the random walk theory of Feller (1968). Markaryan and Rosenberger also derive the exact variance–covariance structure Σ_T and other quantities.

The selection of p has been a source of interest since Efron's original paper, in which he states:

> The value $p = 2/3$, which is the author's personal favourite, will be seen to yield generally good designs and will be featured in the numerical categories.

We have extreme behavior at the endpoints. At $p = 1/2$, we have complete randomization, the most unpredictable randomization procedure, but the maximal imbalance is $n/2$. At $p = 1$, we have the permuted block design with $b = 2$, so that every other treatment assignment is deterministic, but the maximal imbalance for even n is 0. The parameter p, then, represents a trade-off between balance and predictability. Kundt (2007) suggests finding the value of p such that

$$P(|D_n| \geq b) = 1 - p^* \tag{3.18}$$

for a maximum tolerated imbalance b and some probability p^*.

3.6 OTHER BIASED COIN DESIGNS AND GENERALIZATIONS

Soares and Wu (1982) modified Efron's procedure by considering a level of imbalance that would be unacceptable and then imposing a deterministic treatment assignment (by setting $p = 0$) to counter the imbalance. Their design, which they named the *big stick design*, is given by

$$
\begin{aligned}
E(T_j|\mathcal{F}_{j-1}) &= \frac{1}{2}, &&\text{if} \quad |D_{j-1}| < b, \\
&= 0, &&\text{if} \quad D_{j-1} = b, \\
&= 1, &&\text{if} \quad D_{j-1} = -b.
\end{aligned}
$$

The degree of imbalance is given by an *imbalance intolerance* parameter b, which is fixed in advance.

A similar procedure, originally developed by Larry Shaw, was used in the National Cooperative Gallstone Study in which a proportionate degree of imbalance, $D_{j-1}/(j-1)$, was employed in lieu of an absolute difference to define the acceptable degree of imbalance in the aforementioned expression (Lachin *et al.*, 1981).

Chen (1999) introduced a hybrid of the big stick rule and Efron's biased coin design, calling it the *biased coin design with imbalance intolerance*. The rule is given by

$$
\begin{aligned}
E(T_j|\mathcal{F}_{j-1}) &= \frac{1}{2}, &&\text{if} \quad D_{j-1} = 0, \\
&= 0, &&\text{if} \quad D_{j-1} = b, \\
&= 1, &&\text{if} \quad D_{j-1} = -b, \\
&= p, &&\text{if} \quad 0 < D_{j-1} < b, \\
&= 1 - p, &&\text{if} \quad -c < D_{j-1} < 0,
\end{aligned}
$$

and is denoted as BCDII(p). The design induces a random walk on the state space $\{0, \dots, b\}$ with reflecting barriers 0 and b. Asymptotic balancing properties for this stochastic process are derived in Chen (1999).

Baldi Antognini and Giovagnoli (2004) describe a family of designs, which they call *accelerated biased coin designs*, and which have as special cases Efron's biased coin design, the big stick design, and biased coin design with imbalance intolerance. Let F be a function mapping the integers to $[0, 1]$ such that $F(x)$ decreases and $F(-x) = 1 - F(x)$. Then the accelerated biased coin design assigns to treatment A with the rule $E(T_j | F_{j-1}) = F(D_{j-1})$. Within this family of designs, they select a specific function F that has particularly good properties. It is given by

$$
\begin{aligned}
F_a(x) &= \frac{|x|^a}{|x|^a + 1}, & \text{if} \quad x \le -1, \\[2mm]
&= \frac{1}{2}, & \text{if} \quad x = 0, \\[2mm]
&= \frac{1}{|x|^a + 1}, & \text{if} \quad x \ge 1.
\end{aligned}
\tag{3.19}
$$

This procedure is denoted as ABCD(a). The parameter a can be chosen to impact the degree of randomness. If $a = 0$, we have complete randomization. As $a \to \infty$, the design becomes the big stick design with $b = 2$.

3.7 WEI'S URN DESIGN

Using Efron's biased coin design, the bias of the coin, p, is constant, regardless of the degree of imbalance. Wei (1977, 1978a) developed an *adaptive biased coin design*, where the probabilities of assignment adapt according to the degree of imbalance. One convenient model that adapts these probabilities is an urn model. A review of urn randomization is found in Wei and Lachin (1988).

For the urn design, initially an urn contains α balls of each of two types, A and B. When a patient is ready to be randomized, a ball is drawn and *replaced*. If the ball is type A, treatment A is assigned to the patient and β type B balls are added to the urn. If the ball is type B, treatment B is assigned to the patient and β type A balls are added to the urn. In this way, the urn composition is skewed to increase the probability of assignment to the treatment that has been selected least often thus far. As with other designs, the sequence of assignments can be conducted in advance of patient enrollment. The urn design is denoted $UD(\alpha, \beta)$ and has the allocation rule

$$
E(T_j | F_{j-1}) = \frac{\alpha + \beta N_B(j-1)}{2\alpha + \beta(j-1)}, j \ge 2;
$$

$$
E(T_1 | F_0) = \frac{1}{2}.
\tag{3.20}
$$

If $\alpha = 0, \beta = 1$, the first treatment assignment occurs with probability $1/2$. Note that the $UD(\alpha, 0)$ design is complete randomization. For the $UD(0, 1)$, we have the following simple allocation rule:

$$
E(T_j | F_{j-1}) = \frac{N_B(j-1)}{j-1}, j \ge 2.
\tag{3.21}
$$

We can show that the unconditional probability of assignment to A is $1/2$ using induction, using (3.21). First note that $E(T_1) = E\{N_A(1)\} = 1/2$. Assume $E\{N_A(j-1)\} = (j-1)/2$. Then

$$E\{N_A(j)\} = EE\{T_j|F_{j-1}\} + E\{N_A(j-1)\}$$

$$= 1 - \frac{E\{N_A(j-1)\}}{j-1} + \frac{j-1}{2}$$

$$= \frac{j}{2}.$$

Thus, $E(T_j) = 1/2$.

We can examine the transition probabilities for the degree of imbalance, $|D_n|$, of the UD(α, β) as follows. Without loss of generality, assume $N_A(n) - N_B(n) = j$, for positive integer j. Then, since $N_A(n) + N_B(n) = n$, we have $N_A(n) = (j+n)/2$. So

$$\Pr(|D_{n+1}| = j - 1 | |D_n| = j) = \frac{\text{number of type } B \text{ balls in urn}}{\text{total number of balls in urn}}$$

$$= \frac{\alpha + \beta N_A(n)}{2\alpha + \beta n}$$

$$= \frac{1}{2} + \frac{j\beta}{2(2\alpha + n\beta)},$$

$$\Pr(|D_{n+1}| = j + 1 | |D_n| = j) = \frac{1}{2} - \frac{j\beta}{2(2\alpha + n\beta)},$$

$$\Pr(|D_{n+1}| = 1 | |D_n| = 0) = 1.$$

So, asymptotically, UD(α, β) tends to complete randomization. Wei (1977) uses the recursive formula

$$\Pr(|D_{n+1}| = j)$$

$$= \Pr(|D_{n+1}| = j | |D_n| = j - 1) \Pr(|D_n| = j - 1)$$

$$+ \Pr(|D_{n+1}| = j | |D_n| = j + 1) \Pr(|D_n| = j + 1) \qquad (3.22)$$

to find the unconditional distribution of $|D_n|$. Wei (1977, p. 384) tabulates these values for $j \le 10$.

Wei (1978b) shows that $N_A(n)$ is asymptotically normal with mean $n/2$ and asymptotic variance $n/12$, provided $3\beta > \alpha$. Consequently, D_n is asymptotically normal with mean 0 and asymptotic variance $n/3$. We can then compute, for integer r,

$$\Pr(|D_n| > r) \cong 2\left\{1 - \Phi\left(r\sqrt{\frac{3}{n}}\right)\right\}. \qquad (3.23)$$

This can be compared directly with equation (3.1) for complete randomization. In Table 3.3, we evaluate percentiles of the asymptotic imbalance distribution for the

Table 3.3 Percentiles of the distribution of $|D_n|$ for Wei's urn design.

n	0.33	0.25	0.10	0.05	0.025
50	4.0	4.7	6.7	8.0	9.2
100	5.6	6.6	9.5	11.3	12.9
200	8.0	9.4	13.4	16.0	18.3
400	11.2	13.3	19.0	22.6	25.9
800	15.9	18.8	26.9	32.0	36.6

urn design using equation (3.23). We see that, for the UD(0, 1), the urn design has a lower probability of imbalance than complete randomization asymptotically (the entries in Table 3.1 are simply divided by $\sqrt{3}$).

3.8 OTHER URN MODELS AND GENERALIZATIONS

Other urn designs have been proposed in the literature for use in randomized clinical trials. Several will be discussed in the context of response-adaptive randomization in later chapters. Another restricted randomization design for achieving balance is the *Ehrenfest urn model* proposed by Chen (2000). In this urn, b (even) balls are arranged in two urns, labeled A and B, with $b/2$ balls in each urn. The b balls are equally likely to be drawn. One draws one of the b balls at random. If the ball came from urn A, treatment A is assigned and the ball is replaced in urn B. If it came from urn B, treatment B is assigned and it is replaced in urn A. Balancing is achieved because it is more likely to draw a ball from the urn that contains more balls, thus reducing the composition of that urn by 1. Unlike Wei's urn, the Ehrenfest urn maintains a constant number of balls.

The Ehrenfest urn was originally proposed in physics to obtain equilibrium between two isolated bodies. The urn induces a Markov chain on the state space $\{0, \ldots, b\}$ with reflecting barriers at 0 and b and, as such, is directly comparable to Efron's biased coin design with imbalance control described in the previous section. Chen (2000) does a comparison in terms of asymptotic balancing properties and finds that the Ehrenfest urn is more effective than the biased coin design with imbalance control (where b is the same value in both) when, for the biased coin design, $1/b < p < 1/2$. When $0 < p < 1/b$, the biased coin design has better balancing properties.

Wei (1978a) proposed a generalization of his urn model, by defining a function p, where p is a nonincreasing function with continuous derivatives, satisfying $p(x) + p(-x) = 1$. The allocation rule is given by

$$E(T_j|\mathcal{F}_{j-1}) = p\left(\frac{D_{j-1}}{j-1}\right). \tag{3.24}$$

Smith (1984a) proposed a class of designs depending on a positive parameter ρ, given by

$$E(T_j|\mathcal{F}_{j-1}) = \frac{N_B(j)^\rho}{N_A(j)^\rho + N_B(j)^\rho}, j = 2, \dots, n,$$ (3.25)

where $E(T_1) = 1/2$. This rule corresponds to

$$p(x) = \frac{(1-x)^\rho}{(1+x)^\rho + (1-x)^\rho}$$

in equation (3.24). If $\rho = 1$, we have the UD(0, 1). If $\rho = 0$, we have complete randomization. Smith favors the design with $\rho = 5$. Wei (1978b) and Smith (1984b) showed that for procedure (3.25), D_n is asymptotically normal with mean 0 and variance $n/(1 + 2\rho)$.

Wei's design differs from the accelerated biased coin design in the dependence of p on D_n/n instead of D_n. As Baldi Antognini and Giovagnoli (2004) point out, the allocation rule does not distinguish between $n = 5$ and $|D_n| = 1$, in which the design is balanced as much as possible, and $n = 10$ and $|D_n| = 2$, which is an unbalanced design. For direct comparison of the accelerated biased coin design in (3.19) and Smith's design in (3.25), Baldi Antognini and Giovagnoli (2004) suggest setting $a = 2\rho$.

3.9 COMPARISON OF BALANCING PROPERTIES

Figure 3.2 shows the possible paths that six different randomization procedures take for $n = 12$. Efron's biased coin design (a) can take all 2^n possible paths, although extreme paths have very small probability. The big stick design (b) and maximal procedure (c) restrict the reference set of possible sequences by imposing a maximum imbalance bound ($b = 3$ here). The maximal procedure further is restricted by being a forced balance procedure. Each sequence in the maximal procedure is equiprobable. Hadamard randomization (d) reduces the reference set to 22 sequences, each equiprobable. It is difficult to see the difference in the graph from the maximal procedure, but it is not always possible in Hadamard randomization to go up and down from a node, as it is for the maximal procedure. Finally, the permuted block design (e, f) restricts the reference set by imposing balance at 4, 8, and 12 when $m = 4$, and 6 and 12 when $m = 6$.

Table 3.4 gives a simulation comparison of eight adaptive restricted randomization procedures used for balance: complete randomization, Wei's UD(0, 1), Smith's design with $\rho = 5$, Efron's biased coin design ($p = 2/3$), the big stick design and biased coin design with imbalance intolerance (both with $b = 3$), the accelerated biased coin design with $a = 10$, Hadamard randomization using the matrix in (3.13), and the random block design with $B_{max} = 3, 10$. In each case, $n = 50$. Such simulations can be done very quickly in SAS or R, and we present the SAS code for the simulation of Efron's biased coin design in the Appendix in Section 3.14.

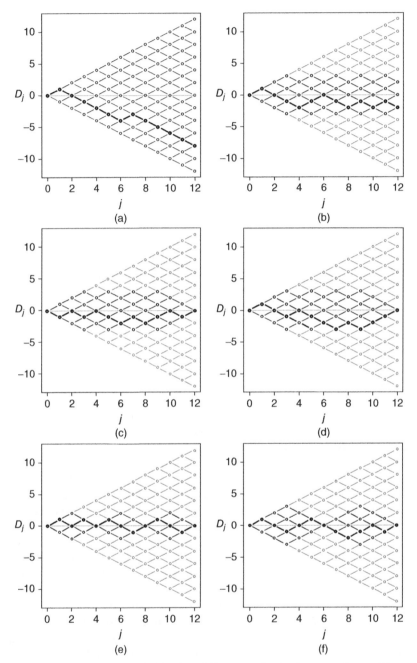

Fig. 3.2 *Possible paths (light bold) of six different randomization procedures, with a particular path outlined in heavy bold. (a) Efron's BCD ($p = \frac{2}{3}$). (b) Big stick design ($b = 3$). (c) Maximal procedure ($b = 3$). (d) Hadamard randomization. (e) Permuted block design ($m = 4$). (f) Permuted block design ($m = 6$).*

Table 3.4 *Simulated variance of D_n and expected maximum imbalance for eight different randomization procedures, $n = 50$, based on $100,000$ replications.*

Randomization Procedure	Var(D_n)	$E(\max_{1 \leq j \leq n} |D_j|)$
Complete	49.92	8.88
UD(0, 1)	16.58	5.83
Smith ($\rho = 5$)	4.69	3.77
BCD(2/3)	4.36	4.28
Big stick ($b = 3$)	2.66	3.00
BCDII(2/3, 3)	1.70	2.94
ABCD(10)	2.01	2.01
Hadamard	1.90	3.23
Random block design ($B_{max} = 3$)	0.75	2.34
Random block design ($B_{max} = 10$)	2.43	3.75

The first three procedures can be thought of as Smith's design with $\rho = 0, 1, 5$, and one sees that the variance is very close to the asymptotic variance of $n/(1 + 2\rho)$, which takes values 50, 16.67, and 4.54, respectively. Only when ρ increases to 5 do we see variability of D_n at the level of Efron's biased coin design and its variants. Efron's, which has no restrictions on the maximum imbalance, has larger variability than the procedures with a maximum imbalance restriction. As a maximum imbalance restriction is imposed, the variability becomes smaller, as can be seen for the big stick design ($p = 1/2$) and its variant, the biased coin design with imbalance intolerance ($p = 2/3$). The accelerated biased coin design has less variability than Efron's. Hadamard randomization in blocks of size 12 leads to a final imbalance only in the terminal incomplete block with two patients. The random block design can only exhibit an imbalance when the last block is unfilled. When the possible block sizes are 2, 4, and 6, the variability is very small, and the maximum possible imbalance is 3, for direct comparison with the big stick design. Increasing B_{max} to 10 increases the variance more than threefold, but the variability is still comparable to biased coin designs. Terminal balance may not be as important as avoiding large imbalances throughout the trial. Consequently, one measure of interest may be the average maximum imbalance, $E(\max|D_j|)$. For instance, the random allocation rule forces $D_n = 0$, but the simulated average maximum imbalance is 3.38, which is much smaller than for the truncated binomial design with average maximum imbalance 8.49 (data not shown). The designs with imbalance intolerance have average maximum imbalance close to b. The effect of B_{max} is seen for the random block design, in that a larger B_{max} leads to a higher maximum imbalance. The maximum imbalance for $B_{max} = 10$ is similar to Smith's design and to Hadamard randomization. Using that metric, the random block design does not perform as well as the biased coin designs and their variants unless B_{max} is small.

In conclusion, if the only consideration is the variability of D_n, the random block design or the biased coin design and their variants are the best procedures. However, as we shall see in future chapters, there are considerations other than balance.

3.10   RESTRICTED RANDOMIZATION FOR UNBALANCED ALLOCATION

Some statisticians (e.g., Peto, 1978) have advocated that, under certain conditions, clinical trials should randomize with fixed unequal allocation probabilities. For example, one might allocate in a $2:1$ ratio of intervention to control. Sometimes, favoring the experimental therapy is warranted in trials of potentially great public health benefit, such as when testing a new AIDS therapy, where patients may be reluctant to have only a 50 percent chance of receiving the new therapy. Such unequal allocation procedures can improve recruitment. One must remember that the experimental therapy may also be harmful, and hence, unequal allocation could subject more patients to a harmful therapy. There are also cases where widespread knowledge about the control therapy exists and more understanding is needed about the experimental therapy. Although the study may lose some sensitivity, there may be gain in terms of information about the toxicity and patient responses to the experimental therapy. This argument was used to justify $2:1$ allocation in an oncology trial (Cocconi *et al.*, 1994). Such decisions could be controversial and should be made in the context of careful power assessments.

In some cases, an optimal allocation ratio different from $1:1$ will maximize power (see, e.g., Problem 2.6). This is discussed further in Chapter 10. Another instance where an unbalanced design is statistically desirable is when the principal analysis is a set of multiple comparisons of $K-1$ treatments versus a single control using the Dunnett (1955) procedure. In this case, the *square-root rule* provides an optimal set of allocation ratios $(K-1)^{1/2}:1:1:\ \ldots\ :1$. For three treatments, the allocation proportions are $(0.414, 0.293, 0.293)$. Rarely, however, is the subset of $K-1$ comparisons employed as the principal analysis in lieu of an overall test of equality of treatment means.

Generally, an unbalanced design, if employed, is justified on the grounds of ethics or cost. In later chapters, we describe the use of response-adaptive randomization, which dynamically alters the allocation probabilities to reflect the accruing data on the trial, putting more patients on the treatment performing better thus far. While different from fixed unbalanced allocation, where the allocation probabilities are determined in advance of the trial, the rationale for response-adaptive randomization is much the same.

In practice, forced balance procedures can be altered to produce a fixed unbalanced allocation. For complete randomization and the truncated binomial design, a biased coin can be tossed with the desired allocation proportion for treatment A. For the random allocation rule, the urn initially contains the desired proportion of balls of each treatment and they are sequentially drawn without replacement.

Suppose the desired allocation is $a:b$, where a and b are integers with no common divisors other than 1, and let $s = a + b$. Then a permuted block design or a random block design can be implemented, provided the block sizes are multiples of s. For small s, this is not problematic, but for larger s, this becomes impractical. For example, a desired $4:7$ allocation would require block sizes of 11, 22, and so on, which is not very practical when implementing a random block design.

Kuznetsova and Tymofyeyev (2013) describe a *brick tunnel* randomization procedure that is feasible for any value of s and limits the maximum imbalance within each block. Consider a block of size s, and we desire to get from the point $(0,0)$ to (a, b), where we move one unit to the right when treatment A is assigned and one unit up when treatment B is assigned. There are $s - 2$ multiples of $1/a$ or $1/b$ in $(0, 1)$. Let $r_i, i = 1, \ldots, s - 2$ be these multiples, and let $r_0 = 0$. The brick tunnel establishes $S - 1$ squares with lower left corners $(\lfloor ar_{i-1} \rfloor, \lfloor br_{i-1} \rfloor), i = 1, \ldots, S - 1$, where $\lfloor \cdot \rfloor$ is the floor function. The lower right and upper left corners of the squares are then uniquely defined by

$$X_{i1} = (\lfloor ar_{i-1} \rfloor + 1, \lfloor br_{i-1} \rfloor);$$
$$X_{i2} = (\lfloor ar_{i-1} \rfloor, \lfloor br_{i-1} \rfloor + 1), i = 1, \ldots, S - 1,$$

respectively. The probability that X_{11} is visited is $p_{11} = a/s$ and the probability that X_{12} is visited is $1 - p_{11} = b/s$. We can compute the remaining probabilities, p_{i1}, of visiting X_{i1} recursively by

$$p_{i+1,1} = p_{i1} - \frac{b}{s}, \text{ if } ar_i \text{ is an integer};$$

$$= p_{i1} + \frac{a}{s}, \text{ otherwise}, \tag{3.26}$$

and $p_{i+1,2} = 1 - p_{i+1,1}$. Then the allocation rule is given conditional on the corner X_{ij} occupied at the ith allocation using the equations:

$$E(T_1) = \frac{a}{s};$$

$$E(T_{i+1}|X_{i1}) = 1 - \frac{b}{(sp_{i1})}, \text{ if } ar_i \text{ is an integer};$$

$$= 0, \text{ otherwise};$$

$$E(T_{i+1}|X_{i2}) = 1, \text{ if } ar_i \text{ is an integer};$$

$$= \frac{a}{(sp_{i2})}, \text{ otherwise}.$$

For example, Figure 3.3 shows the brick tunnel for the allocation ratio $4 : 3$. Here $s = 7$ and the set of unique multiples is $r_0 = 0$, $r_1 = 1/4$, $r_2 = 1/3$, $r_3 = 1/2$, $r_4 = 2/3$, and $r_5 = 3/4$. The brick tunnel establishes six squares with lower left corners $(0,0)$, $(1,0)$, $(1,1)$, $(2,1)$, $(2,2)$, and $(3,2)$. Using (3.26), we can compute $p_{11} = 4/7$, $p_{21} = 1/7$, $p_{31} = 5/7$, $p_{41} = 2/7$, $p_{51} = 6/7$, $p_{61} = 3/7$. This leads to the allocation probabilities (associated with each edge): $E(T_1) = 4/7$, $E(T_2|X_{11}) = 1/4$, $E(T_3|X_{21}) = E(T_5|X_{41}) = 0$, $E(T_4|X_{31}) = 2/5$, $E(T_6|X_{51}) = 1/2$, $E(T_2|X_{12}) = E(T_4|X_{32}) = E(T_6|X_{52}) = 1$, $E(T_3|X_{22}) = 2/3$, $E(T_5|X_{42}) = 4/5$. While a permuted block design with $b = 7$ filled with a random allocation rule would lead to 35 equiprobable sequences, the brick tunnel design leads to only 21 sequences,

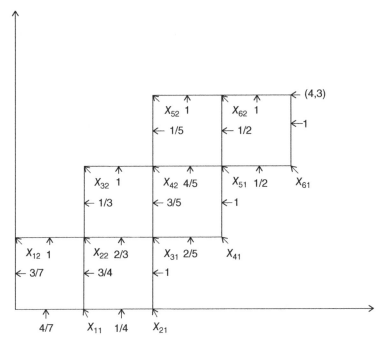

Fig. 3.3   *The brick tunnel for 4 : 3 allocation*

but not equiprobable, and eliminates extreme sequences such as *AAAABBB*. For example, the sequence *AABABAB* has probability $1/35$.

Kuznetsova and Tymofyeyev (2013) show that $E(T_j) = \sum_{l=1}^{L} I(T_{jl} = 1)p_l = a/s$, using the notation of (3.4), a property that they refer to as "preserving the allocation ratio at each step." One can think of the brick tunnel procedure as a proper extension of the maximal procedure for unbalanced allocation, because a naive extension of the maximal procedure with $b = 2$ would also yield 21 sequences, but each would be equiprobable, and hence would not preserve the allocation ratio at each step.

For Efron's biased coin design, Kuznetsova and Tymofyeyev's (2011) suggest randomizing using s "fake" treatments, on which a biased coin for equal allocation is used; the fake treatments are then mapped to the actual treatments. At each randomization, of these s fake treatments, $w < s$ of them are in a "preferred" set treatments that, if assigned, would reduce the range of the imbalance of the trial thus far. The biased coin parameter $p > 1/2$ is divided according to the number of preferred treatments, so that they are allocated with probability p/w, and the non-preferred treatments are allocated with probability $(1 - p)/(s - w)$. If all treatments are balanced, each is assigned with probability $1/s$. Then the treatment allocation is mapped from the fake treatments to the actual treatments by combining the first a fake treatments for treatment A and the remaining $b = s - a$ fake treatments for treatment B.

3.11   *K* > 2 TREATMENTS

For clinical trials of more than two treatments, most of these randomization procedures generalize quite readily. Complete randomization becomes a simple multinomial probability generator with K equally likely outcomes. The random allocation rule can be thought of as an urn with n/K balls representing each treatment. Truncated binomial randomization becomes a multistage process whereby K-treatment complete randomization is used, and each treatment is subsequently dropped when the n/Kth patient is assigned to that treatment, until only one treatment is left. All subsequent patients are then assigned to that treatment.

Kuznetsova and Tymofyeyev (2010) extend brick tunnel randomization to multiple treatments, but instead of operating on squares, they now operate through K dimensional cubes. The name brick tunnel arises from the image of each cube being a brick, and the desired allocation ray piercing the brick enters a "brick tunnel." If $C_1 : C_2 : \ldots : C_K$ is the desired allocation, then a point $U = (C_1 u, C_2 u, \ldots, C_K u), 0 \le u \le 1$ on the allocation ray belongs to the unitary cube on a K-dimensional grid with sides $(\lfloor C_1 u \rfloor, \lfloor C_1 u \rfloor + 1), (\lfloor C_2 u \rfloor, \lfloor C_2 u \rfloor + 1), \ldots, (\lfloor C_K u \rfloor, \lfloor C_K u \rfloor + 1)$. As in the two treatment case, probabilities of visiting corners of the cubes are computed, and then the allocation probabilities are computed conditional on being on a particular corner. In the two treatment cases, this procedure allows only those sequences that satisfy the smallest possible imbalance threshold. Kuznetsova and Tymofyeyev (2013) extend this method by allowing a prespecified maximum imbalance threshold in the K treatment case.

Efron's biased coin design does not have a unique or obvious generalization. Atkinson (2014) suggests ordering the treatments from the most allocated to the least allocated, labeled $i = K, K - 1, \ldots, 1$ (if there are no ties). Then the probability of assigning treatment i is given by $2(K + 1 - i)/K(K + 1)$. Ties can be accommodated by averaging the probabilities over the sets of tied treatments.

Kuznetsova and Tymofyeyev's (2011) suggestion of using "fake" treatments for unequal allocation generalizes easily to $K > 2$ treatments with unequal allocation. For example, if $K = 3$ and the desired allocation is $2 : 1 : 1$, then there would be $s = 4$ fake treatments, and Efron's biased coin would be applied to the fake treatments for each patient, with the bias weights being split among the "preferred" arms that decrease the total imbalance. Then fake treatments 1 and 2 would be mapped to treatment 1, fake treatment 3 to treatment 2, and fake treatment 4 to treatment 3.

This rule for the division of the bias p across treatments was initially proposed by Pocock and Simon (1975). However, note that this rule may lead to the least represented treatment having the lowest individual probability of being augmented. For example, if the allocation is $2 : 3 : 2 : 2$ for four fake treatments and $p = 2/3$, then $w = 3$, and the next allocation will be with probabilities $2/9 : 1/3 : 2/9 : 2/9$, so that fake treatment 2 would have the highest probability of augmentation. Suppose we wish to map these four fake treatments to three treatments in a $2 : 1 : 1$ ratio. Then the current allocation is $5 : 2 : 2$ among the actual treatment arms.

The procedure would then augment the first treatment with probability 5/9, leading to a higher probability of attaining a 3 : 1 : 1 ratio.

Wei's urn design admits an easier generalization. For the UD(α, β) design, the urn contains α balls representing each treatment initially. Then β balls are added for each other treatment after an assignment is made. For the UD(0, 1), the probability that the jth assignment is to treatment i, given the previous $j - 1$ assignments, is given by

$$\frac{j - 1 - N_i(j - 1)}{(j - 1)(K - 1)}. \tag{3.27}$$

This reduces to (3.21) when $K = 2$.

Wei, Smythe, and Smith (1986) describe a wide class of designs for K treatments that are an extension of (3.24). Suppose the desired allocation proportions for the K groups is $\boldsymbol{\xi} = (\xi_1, \ldots, \xi_K)$ (with all $\xi_i = 1/K$ equal unless one wishes unbalanced allocation; see Section 3.10).

Let $N(j) = (N_1(j), \ldots, N_{K-1}(j))$ be the number of patients assigned to treatment $i = 1, \ldots, K - 1$ after j patients have been assigned. Define $\boldsymbol{p} = (p_1(\boldsymbol{x}), \ldots, p_{K-1}(\boldsymbol{x}))$ as continuous functions of some $(K - 1) \times 1$ vector $\boldsymbol{x}$ (typically $p_1, \ldots, p_K$ will be a function of $\boldsymbol{x} = N(j - 1)/(j - 1)$), where $p_i(\boldsymbol{x})$ is the probability that patient j will be assigned to treatment i and $p_K = 1 - \sum_{k=1}^{K-1} p_k$. Then $p_1(\boldsymbol{x}), \ldots, p_K(\boldsymbol{x})$ are assumed to satisfy the following relationship:

$$\text{if } x_i \geq \xi_i, \text{ then } p_i(\boldsymbol{x}) \leq \xi_i, j = 1, \ldots, K. \tag{3.28}$$

For some special cases of this general procedure, consider first complete randomization. Here $p_i(\boldsymbol{x}) = 1/K$ for all i, so that future assignments are independent of previous assignments. For the generalization of Wei's urn in (3.27), we have $p_i(\boldsymbol{x}) = (1 - x_i)/(K - 1)$. Note that Efron's biased coin design is not a special case of (3.28) because $\boldsymbol{p}$ is not continuous for each i.

We have the following important limiting result from Wei, Smythe, and Smith (1986):

$$p_i\left(\frac{N(j - 1)}{j - 1}\right) \rightarrow \xi_i, \tag{3.29}$$

in probability, as $j \rightarrow \infty$, for $i = 1, \ldots, K$. So, this general rule for K treatments tends asymptotically to the desired allocation.

3.12 PROBLEMS

3.1 Using each of the following procedures, generate a separate randomization sequence for 50 random allocations to two groups:

(i) complete randomization;
(ii) random allocation rule;
(iii) truncated binomial design;

(iv) permuted block design with $M = 5$;
(v) random block design with $B_{\max} = 3$;
(vi) Efron's biased coin with $p = 2/3$;
(vii) big stick design with $b = 3$;
(viii) accelerated biased coin design with $a = 10$;
(ix) Wei's urn design with $\alpha = 0$ and $\beta = 1$;
(x) Smith's procedure with $\rho = 5$.

Provide the sequence and a copy of your program for each of the ten procedures.

3.2 After 10 patients have been randomized, $D_{10} = -2$. Compute the probability that patient 11 will be assigned to treatment A for the following randomization procedures:

(i) complete randomization;
(ii) random allocation rule;
(iii) truncated binomial design;
(iv) permuted block design with $m = 6$;
(v) Efron's biased coin with $p = 2/3$;
(vi) big stick design with $b = 3$;
(vii) accelerated biased coin design with $a = 10$;
(viii) Wei's urn design with $\alpha = 0$ and $\beta = 1$;
(ix) Smith's procedure with $\rho = 5$.

3.3 Repeat Figure 3.1 for the standard Z-test of a difference of two proportions.

3.4 Analyze the balancing property of complete randomization and Wei's urn design ($\alpha = 0, \beta = 1$) theoretically using the normal approximations given in equations (3.1) and (3.23). For imbalances $\Pr(|D_n|/n > r/n)$, and $r/n = 0.05, 0.10, 0.20$, graph the probability of imbalance versus sample size (for $n = 1$–100) with graphics software. Superimpose the three curves (for the three values of r/n) for each procedure. Interpret and compare the results with Table 3.4.

3.5 Derive equations (3.7) and (3.8).

3.6 Consider a permuted block design with block size $m = 12$, filled using the random allocation rule, and filled using the Hadamard randomization defined by (3.13). Show the following (Bailey and Nelson, 2003):

a. If the trial is stopped in the final block after 2 or 10 patients, the probability of perfect balance is $6/11$ for both procedures.
b. If the trial is stopped in the final block after 4 or 8 patients, $P(|D_n| = 4) = 0.61$ using the random allocation rule and 0 using Hadamard randomization.
c. Suppose a trial with target size 50 is stopped after 42 patients. If the blocks are filled using the random allocation rule, $P(|D_n| = 0) = 0.44$, but using Hadamard randomization, it is reduced to 0.18. What is the maximum imbalance for each procedure?
d. Draw conclusions.

3.7 Show that $N_{\text{MP,b,n}}/N_{\text{PBD,b,n}}$ increases in n. (*Hint*: What is the relationship between $N_{\text{PBD,b,n}}$ and $N_{\text{PBD,b,n}+2b}$?) (Berger, Ivanova, and Knoll, 2003)

3.8 Derive the solution to the steady-state equations for Efron's biased coin design, given in equations (3.16).

3.9 Derive the steady-state variance of D_n for the $BCD(p)$ in equation (3.17). How does this result compare to the value given in Table 3.4?

3.10 Show that $p = 0.645$ is a solution to (3.18) for $b = 6$ and $p^* = 0.95$, using the steady-state probabilities from (3.16). (Kundt, 2007)

3.11 Consider brick tunnel randomization with $3:2$ allocation (Kuznetsova and Tymofyeyev, 2010):

a. Redraw Figure 3.3 with the transition probabilities.
b. Enumerate the eight possible sequences and give their associated probabilities.
c. Show that $E(T_i) = 3/7, i = 1, \ldots, 8$, for the brick tunnel, but that if each sequence were equiprobable, this property would fail.

3.13 REFERENCES

ANSCOMBE, F. J. (1948). The validity of comparative experiments. *Journal of the Royal Statistical Society A* **111** 181–211 (with discussion).

ATKINSON, A. C. (2014). Selecting a biased-coin design. *Statistical Science* **29** 144–163.

BAILEY, R. A. and NELSON, P. R. (2003). Hadamard randomization: a valid restriction of random permuted blocks. *Biometrical Journal* **45** 554–560.

BALDI ANTOGNINI, A. and GIOVAGNOLI, A. (2004). A new 'biased coin design' for the sequential allocation of two treatments. *Journal of the Royal Statistical Society C* **53** 651–664.

BERGER, V. W., IVANOVA, A., and KNOLL, M. D. (2003). Minimizing predictability while retaining balance through the use of less restrictive randomization procedures. *Statistics in Medicine* **22** 3017–3028.

BLACKWELL, D. and HODGES, J. L. (1957). Design for the control of selection bias. *Annals of Mathematical Statistics* **28** 449–460.

CHEN, Y.-P. (1999). Biased coin design with imbalance intolerance. *Communications and Statistics–Stochastic Models* **15** 953–975.

CHEN, Y.-P. (2000). Which design is better? Ehrenfest urn versus biased coin. *Advances in Applied Probability* **32** 738–749.

COCCONI, G., BELLA, M., ZIRONI, S., ALGERI, R., DI COSTANZO, F., DE LISI, V., LUPPI, G., MAZZOCCHI, B., RODINO, C., and SOLDANI, M. (1994). Fluorouracil, doxorubicin, and mitomycin combination versus PELF chemotherapy in advanced gastric cancer: a prospective randomized trial of the Italian Oncology Group for Clinical Research. *Journal of Clinical Oncology* **12** 2687–2693.

DUNNETT, C. W. (1955). A multiple comparison procedure for comparing several treatments with a control. *Journal of the American Statistical Association* **50** 1096–1121.

EFRON, B. (1971). Forcing a sequential experiment to be balanced. *Biometrika* **58** 403–417.

FELLER, W. (1968). *An Introduction to Probability Theory and its Applications.* Vol. **I**. John Wiley & Sons, Inc., New York.

FRIEDMAN, L. M., FURBERG, C. D., and DEMETS, D. L. (1981). *Fundamentals of Clinical Trials.* Wright PSG, Boston, MA.

HAINES, L. M. (2013). A random walk approach for deriving exact expressions for probabilities associated with Efron's biased coin design. *South African Journal of Statistics* **47** 123–125.

HEDAYAT, A. S., SLOANE, N. J. A., and STUFKEN, J. (1999). *Orthogonal Arrays.* Springer-Verlag, New York.

HEUSSEN, N. (2004). *Der Einfluss der Randomisierung in Blöcken zufälliger Länge auf die Auswertung klinischer Studien mittels Randomisationstest.* Aachen: RWTH Aachen University (doctoral dissertation).

HU, F. and ROSENBERGER, W. F. (2006). *The Theory of Response-Adaptive Randomization in Clinical Trials.* John Wiley & Sons, Inc., New York.

KARLIN, S. and TAYLOR, H. M. (1975). *A First Course in Stochastic Processes.* Academic Press, Boston, MA.

KUNDT, G. (2007). A new proposal for setting parameter values in restricted randomization methods. *Methods of Information in Medicine* **46** 440–449.

KUZNETSOVA, O. and TYMOFYEYEV, Y. (2010). Brick tunnel randomization for unequal allocation to two or more treatment groups. *Statistics in Medicine* **30** 812–824.

KUZNETSOVA, O. and TYMOFYEYEV, Y. (2011). Preserving the allocation ratio at every allocation with biased coin randomization and minimization in studies with unequal allocation. *Statistics in Medicine* **31** 701–723.

KUZNETSOVA, O. and TYMOFYEYEV, Y. (2013). Wide brick tunnel randomization–an unequal allocation procedure that limits the imbalance in treatment totals. *Statistics in Medicine* **33** 1514–1530.

LACHIN, J. M., MARKS, J., SCHOENFIELD L. J., and the Protocol Committee and the NCGS Group. (1981). Design and methodological considerations in the National Cooperative Gallstone Study: a multi-center clinical trial. *Controlled Clinical Trials* **2** 177–230.

LUMLEY, T. (2012). SURVEY: Analysis of complex survey samples. R, Version 3.28-2.

MARKARYAN, T. and ROSENBERGER, W. F. (2010). Exact properties of Efron's biased coin randomization procedure. *Annals of Statistics* **38** 1546–1567.

PETO, R. (1978). Clinical trial methodology. *Biomedicine* **28** 24–36.

POCOCK, S. J. and SIMON, R. (1975). Sequential treatment assignment with balancing for prognostic factors in the controlled clinical trial. *Biometrics* **31** 103–115.

ROSS, S. M. (1983). *Stochastic Processes.* John Wiley & Sons, Inc., New York.

SALAMA, I., IVANOVA, A., and QAQISH, B. (2008). Efficient generation of constrained block allocation sequences. *Statistics in Medicine* **27** 1421–1428.

SCHULTZ, K. F. and GRIMES, D. A. (2002). Unequal group sizes in randomised clinical trials: guarding against guessing. *The Lancet* **359** 966–970.

SMITH, R. L. (1984a). Sequential treatment allocation using biased coin designs. *Journal of the Royal Statistical Society B* **46** 519–543.

SMITH, R. L. (1984b). Properties of biased coin designs in sequential clinical trials. *Annals of Statistics* **12** 1018–1034.

SOARES, J. F. and WU, C. F. J. (1982). Some restricted randomization rules in sequential designs. *Communications in Statistics – Theory and Methods* **12** 2017–2034.

WEI, L. J. (1977). A class of designs for sequential clinical trials. *Journal of the American Statistical Association* **72** 382–386.

WEI, L. J. (1978a). An application of an urn model to the design of sequential controlled clinical trials. *Journal of the American Statistical Association* **73** 559–563.

WEI, L. J. (1978b). The adaptive biased coin design for sequential experiments. *Annals of Statistics* **6** 92–100.

WEI, L. J. and LACHIN, J. M. (1988). Properties of the urn randomization in clinical trials. *Controlled Clinical Trials* **9** 345–364.

WEI, L. J., SMYTHE, R. T., and SMITH, R. L. (1986). K-treatment comparisons with restricted randomization rules in clinical trials. *Annals of Statistics* **14** 265–274.

ZELEN, M. (1974). The randomization and stratification of patients to clinical trials. *Journal of Chronic Diseases* **27** 365–375.

3.14 APPENDIX

The following SAS code simulated the balancing behavior of Efron's biased coin design. The user specified the bias parameter p and the seed for the simulation. The results are given in Table 3.4 for $p = 2/3$.

```
data one;
 n=50;
 do i=1 to 100000;
 d=0;
 maximbal=0;
 p=;
 seed=;
 do j=1 to n;
   if d=0 then proba=1/2;
   if d>0 then proba=1-p;
   if d<0 then proba=p;
   x=ranuni(seed);
   if x < proba then d+1;
   else d=d-1;
 if abs(d) > maximbal then maximbal = abs(d);
  end;
  output;
 end;
 proc means mean var;
  var d maximbal;
run;
```

4

The Effects of Unobserved Covariates

4.1 INTRODUCTION

In Section 1.3, a historic quote from Cornfield gave two important reasons for randomization. Here we examine the first: the probability of an imbalance between treatments with respect to known or unknown covariates tends to zero as n gets large. In Chapters 7 and 9, we will describe randomization techniques to promote balance on known covariates. But the human physiology is so complex that it is simply impossible to identify every covariate that may be related to treatment outcome. One can certainly adjust for any covariates collected during the clinical trial in a post-hoc analysis, but it is likely that some important covariates may not have been collected. Thus, we explore one of the great benefits of randomization: it tends to minimize any incomparability between treatment groups with regard to unknown covariates.

Some statisticians have argued that randomization is unnecessary. However, a simple example, adapted from Berry (1989), demonstrates its benefits. Suppose one wishes to compare the time it takes to drive to work on two different routes. Does it make a difference in our conclusions if we drive five times consecutively on one route and then five times consecutively on the other route or if we flip a coin for 10 days to randomly to select the route? One can quickly respond that, yes, it does matter, if it snowed the first 5 days and there was good weather on the latter 5 days. Intuitively, it would seem that randomizing the route would make it more likely to have similar number of days of bad weather for both routes. This is the concept that randomization tends to balance treatment assignments on unknown covariates that may be related to treatment. This intuition is correct, but only asymptotically. For small samples, there is still a significant probability that an imbalance will result. For the extreme case, randomizing the experiment still leads to a probability that one route is taken only in bad weather of 1/252. Consequently, believers in randomization will talk

Randomization in Clinical Trials: Theory and Practice, Second Edition.
William F. Rosenberger and John M. Lachin.
© 2016 John Wiley & Sons, Inc. Published 2016 by John Wiley & Sons, Inc.

only about "mitigating" covariate imbalances. In fact, everyone who has participated in a randomized clinical trial has witnessed covariate imbalances among treatment groups, even with carefully conducted randomization and relatively large numbers of randomized subjects. It can be presumed that such phenomena provide nonbelievers with some degree of satisfaction.

In this chapter, we explore some of the theoretical properties of randomization in mitigating biases from unknown covariates. We begin with simple probability statements on covariate imbalances and then examine Efron's more sophisticated model for accidental bias.

4.2   A BOUND ON THE PROBABILITY OF A COVARIATE IMBALANCE

The simplest method to analyze the probability of a covariate imbalance is to use Chebyshev's inequality. Suppose $T_1, \ldots, T_n$ is a randomization sequence generated from a forced balance restricted randomization procedure where $E(T_i) = Q, i = 1, \ldots, n$. Let $F_n = \{T_1, \ldots, T_n\}$. Let $n_A = \sum_{i=1}^{n} T_i$ be fixed in advance, and $n_A = Qn$ is an integer for $Q \in (0, 1)$. Suppose Z is some covariate of interest independent of treatment assignments, and $Z_1, \ldots, Z_n$ are independent mean μ_i and variance $\sigma^2, i = 1, \ldots, n$. Then $\bar{Z}_A = \sum_{i=1}^{n} T_i Z_i / (Qn)$ and $\bar{Z}_B = \sum_{i=1}^{n} (1 - T_i) Z_i / ((1 - Q)n)$. A covariate imbalance between treatment groups would be represented by a tangible difference between $\bar{Z}_A$ and $\bar{Z}_B$. Using a conditioning argument, we can then derive the following:

$$
E\{\bar{Z}_A - \bar{Z}_B\} = EE\left\{ \frac{\sum_{i=1}^{n} Z_i T_i}{Qn} - \frac{\sum_{i=1}^{n} Z_i(1 - T_i)}{(1 - Q)n} \middle| F_n \right\}
$$

$$
= \frac{\sum_{i=1}^{n} \mu_i E(T_i)}{Qn} - \frac{\sum_{i=1}^{n} \mu_i E(1 - T_i)}{(1 - Q)n}
$$

$$
= 0 \tag{4.1}
$$

and

$$
\mathrm{Var}\{\bar{Z}_A - \bar{Z}_B\} = E\mathrm{Var}\left\{ \frac{\sum_{i=1}^{n} Z_i T_i}{Qn} - \frac{\sum_{i=1}^{n} Z_i(1 - T_i)}{(1 - Q)n} \middle| F_n \right\}
$$

$$
= \frac{\sigma^2 \sum_{i=1}^{n} E(T_i)}{Q^2 n^2} + \frac{\sigma^2 \sum_{i=1}^{n} E(1 - T_i)}{(1 - Q)^2 n^2}
$$

$$
= \frac{\sigma^2}{Q(1 - Q)n}.
$$

Then by an application of Chebyshev's inequality, for any $\epsilon > 0$,

$$
\mathrm{Pr}\{|\bar{Z}_A - \bar{Z}_B| \geq \epsilon\} \leq \frac{\sigma^2}{\epsilon^2 Q(1 - Q)n}
$$

$$
\to 0 \text{ as } n \to \infty.
$$

It can easily be seen that we can extend this result to heterogeneous variances $\sigma_i^2, i = 1, \ldots, n$, provided $\sum_{i=1}^{n} \sigma_i^2$ is $o(n^2)$.

We can immediately see the problem if balanced, but nonrandom, allocation is employed. Suppose every even patient is assigned to A and every odd patient is assigned to B for n even. The T_i's are deterministic, and equation (4.1) becomes

$$E\{\bar{Z}_A - \bar{Z}_B\} = \frac{\sum_{i=1}^{n/2} \mu_{2i}}{\frac{n}{2}} - \frac{\sum_{i=1}^{n/2} \mu_{2i-1}}{\frac{n}{2}},$$

which is clearly not 0.

Now suppose complete randomization is used, in which case the final numbers assigned to A and B are random variables, given by $N_A(n)$ and $N_B(n) = n - N_A(n)$. Note that $N_A \sim b(n, 1/2)$. Now a formal proof of the Chebyshev bound becomes more challenging. The mean covariate values on each treatment are given by $\bar{Z}_A = \sum_{i=1}^{n} T_i Z_i / N_A(n)$ and $\bar{Z}_B = \sum_{i=1}^{n} (1 - T_i) Z_i / N_B(n)$, respectively. Note that $N_A(n) = 0$ or $N_B(n) = 0$ gives $0/0$, which is undefined, but under complete randomization these events have positive (albeit small) probability. So, formally we have to condition the problem on the set $\{N_A(n), N_B(n) > 0\}$.

Assume the following:

$$(A1) \quad \sum_{i=1}^{n} \mu_i^2 = o(n^2).$$

We use a conditioning argument, conditional on $N_A(n)$, to obtain the mean and variance. We compute $E\{\bar{Z}_A - \bar{Z}_B\}$ via

$$EE\left\{ \frac{\sum_{i=1}^{n} Z_i T_i}{N_A(n)} - \frac{\sum_{i=1}^{n} Z_i (1 - T_i)}{N_B(n)} \middle| N_A(n); N_A(n), N_B(n) > 0 \right\}$$

$$= E\left\{ \frac{\sum_{i=1}^{n} \mu_i E(T_i | N_A(n))}{N_A(n)} - \frac{\sum_{i=1}^{n} \mu_i E(1 - T_i | N_A(n))}{N_B(n)} \middle| N_A(n), N_B(n) > 0 \right\}.$$

(4.2)

Noting that $T_i | N_A(n)$ are now dependent Bernoulli indicators with parameter $N_A(n)/n$, we have $E(T_i | N_A(n)) = N_A(n)/n$ and $E(1 - T_i | N_A(n)) = N_B(n)/n$. Consequently, equation (4.2) is 0.

Similarly, for $\text{Var}\{\bar{Z}_A - \bar{Z}_B\}$, we compute

$$E\text{Var}\left\{ \frac{\sum_{i=1}^{n} Z_i T_i}{N_A(n)} - \frac{\sum_{i=1}^{n} Z_i (1 - T_i)}{N_B(n)} \middle| N_A(n); N_A(n), N_B(n) > 0 \right\}$$

$$= E\left\{ \frac{\sum_{i=1}^{n} \text{Var}(Z_i T_i | N_A(n))}{(N_A(n))^2} + \frac{\sum_{i=1}^{n} \text{Var}(Z_i (1 - T_i) | N_A(n))}{(N_B(n))^2} \middle| N_A(n), N_B(n) > 0 \right\},$$

(4.3)

as the covariance is 0 since $T_i(1 - T_i) = 0$. Since Z_i and T_i are independent,

$$\text{Var}(Z_i T_i | N_A(n)) = (\sigma^2 + \mu_i^2)\text{Var}(T_i | N_A(n)) + \sigma^2 \{E(T_i | N_A(n))\}^2,$$

where $\text{Var}(T_i | N_A(n)) = N_A(n)N_B(n)/n^2$. So $\text{Var}\{\bar{Z}_A - \bar{Z}_B\}$

$$
\begin{aligned}
= \ & \frac{\sigma^2}{n} E\left(\frac{N_B(n)}{N_A(n)} + \frac{N_A(n)}{N_B(n)} + 2 \,\middle|\, N_A(n), N_B(n) > 0\right) \\
& + \frac{\sum_{i=1}^{n}\mu_i^2}{n^2} E\left(\frac{N_B(n)}{N_A(n)} + \frac{N_A(n)}{N_B(n)} \,\middle|\, N_A(n), N_B(n) > 0\right).
\end{aligned}
\tag{4.4}
$$

One can apply a result on the first inverse moment of the positive binomial distribution (2009, Žnidarič) to show that

$$E\left(\frac{N_B(n)}{N_A(n)} \,\middle|\, N_A(n), N_B(n) > 0\right) = 1 + O(n).$$

The Chebyshev bound is therefore

$$
\begin{aligned}
\Pr\{|\bar{Z}_A - \bar{Z}_B| \geq \epsilon\} \leq \ & \frac{4\sigma^2}{\epsilon^2 n} + \frac{2\sum_{i=1}^{n}\mu_i^2}{n^2} \\
& \to 0 \text{ as } n \to \infty,
\end{aligned}
\tag{4.5}
$$

by assumption (A1). For more complicated restricted randomization procedures, such as Efron's biased coin design and its variants, it can be presumed that the Chebyshev bound holds similarly, but the formal proof of this is an open problem.

4.3   SIMULATION RESULTS

To explore the practical susceptibility to covariate imbalances, we simulated the probability of a covariate imbalance for various randomization procedures. For $n = 100$, let $Z_1, \ldots, Z_n$ be covariate values with mean $\bar{Z}_A$ on treatment A and $\bar{Z}_B$ on treatment B. We computed the simulated probabilities $\Pr(|\bar{Z}_A - \bar{Z}_B| > \epsilon)$, where we set $\epsilon = 0.4$. We investigate three different streams of covariates.

1. $Z_1, \ldots, Z_n$ are i.i.d. $N(0, 1)$.
2. $Z_1, \ldots, Z_n$ are subject to a drift over time, ranging linearly on the interval $(-2, 2]$ plus a $N(0, 1)$ random variable.
3. $Z_1, \ldots, Z_n$ are autocorrelated. Random variables $Y_1, \ldots, Y_n$ are generated as i.i.d. $N(0, 1)$, and $Z_j = Y_j + Z_{j-1}, j = 2, .., n, Z_1 = Y_1$.

Table 4.1 Simulated $P(|\bar{Z}_A - \bar{Z}_B| > 0.4)$ for three different types of covariate streams, $n = 50, 100, \ 100,000$ replications.

Procedure	Model (1)		Model (2)		Model (3)	
	$n = 50$	$n = 100$	$n = 50$	$n = 100$	$n = 50$	$n = 100$
CR	0.162	0.047	0.363	0.195	0.312	0.158
RAR	0.157	0.046	0.357	0.190	0.308	0.155
TBD	0.157	0.045	0.541	0.394	0.329	0.170
RBD ($B_{max} = 3$)	0.156	0.045	0.169	0.049	0.246	0.104
RBD ($B_{max} = 10$)	0.157	0.045	0.205	0.059	0.291	0.138
UD(0, 1)	0.159	0.046	0.285	0.131	0.298	0.151
Smith ($\rho = 5$)	0.157	0.045	0.217	0.081	0.273	0.136
BCD (2/3)	0.158	0.045	0.236	0.071	0.291	0.139
Big stick ($b = 3$)	0.157	0.046	0.218	0.061	0.274	0.123
BCDII(2/3, 3)	0.157	0.045	0.191	0.055	0.277	0.128
ABCD(10)	0.157	0.046	0.193	0.055	0.248	0.105

Note that under model (1), a perfectly balanced procedure should lead to $P(|\bar{Z}_A - \bar{Z}_B| > \epsilon) = 0.046$ for $n = 100$ and 0.157 for $n = 50$. Table 4.1 gives the results of our simulation. Each procedure is simulated 100,000 times. For randomization sequences, we used complete randomization (CR), the random allocation rule (RAR), the truncated binomial design (TBD), the random block design (RBD) with $B_{max} = 3, 10$, Wei's urn design UD(0, 1) (UD), Smith's design with $\rho = 5$, Efron's biased coin design ($p = 2/3$), the big stick design and biased coin design with imbalance intolerance (both with $b = 3$), and the accelerated biased coin design with $a = 10$. The SAS code for the simulation of the random block design is given in the Appendix in Section 4.10.

Not surprisingly, under an i.i.d. stream of $N(0, 1)$ variables, all procedures yield simulated results close to the true probability under equal allocation. We note in this case, the Chebyshev bound applies whether or not we used randomization. Models (2) and (3) are the interesting ones. Procedures such as complete randomization give a positive probability of sequences like *AAAAA … BBBBB*, which would be the worst case under a linear time trend, and hence, the more restricted procedures are better. Clearly the truncated binomial design is the worst, and there is a theoretical basis for that, as we will see in Section 4.5. As we saw with the simple balancing properties in Table 3.4, the random block design with $B_{max} = 3$ and the accelerated biased coin design with $a = 10$ are the best procedures with respect to balancing on covariates for both models (2) and (3), with the random block design outperforming the accelerated biased coin design for model (2).

4.4 ACCIDENTAL BIAS

The development in Section 4.2 focuses on the probability of a covariate imbalance, but does not address the effect that such an imbalance may have on the results of the clinical trial. Efron (1971) introduced the term *accidental bias* to describe a measure of the bias in the treatment effect induced by an unobserved covariate. Consider the normal error linear model, from which we will estimate the treatment effect by the ordinary least squares method. Here we modify notation from Chapter 3 slightly. Let $T = (T_1, \dots, T_n)'$ be centered treatment indicators, that is, $T_i = 1$ if treatment A and $T_i = -1$ if treatment B, $i = 1, \dots, n$. We will assume that $E(T) = 0$. (Note that all the randomization procedures in Chapter 3 have this property.) Let $Y = (Y_1, \dots, Y_n)'$ be a vector of responses to treatment. Suppose we fit a standard normal error regression model, where the mean response, conditional on $T = t = (t_1, \dots, t_n)'$, is given by

$$E(Y) = \mu e + \alpha t, \tag{4.6}$$

where $e = (1, 1, \dots, 1)'$. Under equation (4.6), the design matrix is

$$X = [e : t],$$

and hence

$$(X'X)^{-1} = \frac{1}{n^2 - (e't)^2} \begin{bmatrix} n & -e't \\ -e't & n \end{bmatrix},$$

$$X'Y = \begin{bmatrix} e'Y \\ t'Y \end{bmatrix}.$$

Then

$$\hat{\alpha} = \frac{nt'Y - (e't)(e'Y)}{n^2 - (e't)^2}.$$

However, we have ignored a covariate, $z = (z_1, \dots, z_n)'$, that is important in the model. Without loss of generality, assume $z'e = 0$ and $z'z = 1$. Then the correctly specified model is given by

$$E(Y) = \mu e + \alpha t + \beta z. \tag{4.7}$$

Taking the expectation with respect to Y when model equation (4.7) is correct, we obtain

$$E(\hat{\alpha}) = \frac{n(\mu e't + n\alpha + \beta z't) - e't(n\mu + \alpha e't)}{n^2 - (e't)^2}$$

$$= \alpha + \frac{n}{n^2 - (e't)^2} \beta z't.$$

The squared bias term is then given by

$$\{E(\hat{\alpha} - \alpha)\}^2 = \left(\frac{n}{n^2 - (e't)^2} \right)^2 \beta^2 (z't)^2.$$

It is clear that we should desire $e't = 0$ or that the treatment assignments be balanced to minimize accidental bias. If that is accomplished, the degree to which we are subject to accidental bias is controlled by the term $(z't)^2$, which is zero if z is orthogonal to t. Since t is a realization of T, we can obtain the unconditional expectation, by taking $E(z'T)^2$ for a fixed vector z, and we obtain

$$E(z'T)^2 = z'\Sigma_T z,$$

where $\Sigma_T = \text{Var}(T)$. By a result of Rao (1973, p. 62), $z'\Sigma_T z$ cannot exceed the maximum eigenvalue of Σ_T, and the inequality is sharp if the corresponding eigenvector is orthogonal to e. So, Efron uses the maximum eigenvalue of Σ_T as a criterion to define the degree to which a randomization procedure is subject to accidental bias. This yields a minimax criterion when used as the basis for determining a randomization procedure $T_1, \ldots, T_n$ that minimizes the maximum possible value of $z'\Sigma_T z$.

4.5 MAXIMUM EIGENVALUE OF Σ_T

Note that, for complete randomization, $\Sigma_T = I$, where I is the identity matrix, and hence the maximum eigenvalue is 1. This is the smallest possible value for the maximum eigenvalue of Σ_T (see Problem 4.3). For restricted randomization, it is usually not a trivial exercise to derive the variance–covariance structure of the treatment assignments. In Chapter 3, we were able to derive the covariances for the random allocation rule and the truncated binomial design. We will now explore the behavior of the maximum eigenvalue, denoted as $\lambda_{\max}$, for these two designs.

For the random allocation rule, we have from equation (3.5) that $\text{cov}(T_i, T_j) = -1/(n-1)$ (the factor 4 in the denominator disappears when the treatment assignments are 1 and -1), so that Σ_T is of the form $aI + bJ$, where $a = 1 + 1/(n-1)$ and $b = -1/(n-1)$ (J is a matrix of 1's). Then we can derive

$$\lambda_{\max} = 1 + \frac{1}{n-1}.$$

So as $n \to \infty$, the accidental bias becomes negligibly small compared to complete randomization. From this result we also obtain the maximum eigenvalue for the random allocation rule within a permuted block of size $m = n/M$ as

$$\lambda_{\max} = 1 + \frac{1}{m-1}$$

(Problem 4.4).

The truncated binomial design is far more complicated. From equation (3.11), we have $\text{cov}(T_i, T_j) = \text{Pr}(\tau < i)$, and the probability statement involves the truncated negative binomial distribution in equation (3.12). For n even, define

$H(k) = \sum_{l=1}^{k} \Pr(\tau = n/2 + l - 1)$. Since $\text{cov}(T_i, T_j) = 0$ if $i < (n+2)/2$, we can write the variance–covariance matrix in a block structure:

$$\Sigma_T = \begin{bmatrix} I & 0 \\ 0 & C \end{bmatrix},$$

where C is an $n/2 \times n/2$ matrix with elements $(1 - \delta_{ij})H(\min(i,j)) + \delta_{ij}$ (δ_{ij} is the Kronecker delta). Rosenberger and Rukhin (2002) then prove that

$$\sqrt{\pi \frac{n}{3}} \leq \lambda_{\max} \leq \sqrt{\frac{n}{2}},$$

so that $\lambda_{\max}$ grows like $n^{1/2}$. It is the determinism of the tail sequence that induces correlations that drive the accidental bias to infinity as $n \to \infty$. Consequently, the truncated binomial design is not protective against accidental bias.

4.6 ACCIDENTAL BIAS FOR BIASED COIN DESIGNS

Rather than directly examining Σ_T (which he acknowledges is difficult), Efron (1971) looks at the much simpler process $T_1, T_2, T_3, \ldots, T_n$, assuming that it is stationary, and aims at finding the asymptotic covariance structure of the process. He then shows that the asymptotic maximum eigenvalue of the covariance vector $(T_{h+1}, \ldots, T_{h+N})$ as $h \to \infty$, λ_N, is increasing in N and has a finite limit. Based on numerical evidence, Efron conjectures that $\lim_{N\to\infty} \lambda_N = 1 + (p - q)^2$. This was later proved by Steele (1980).

However, Smith (1984) shows by counterexample that Efron's solution may be unsatisfactory when there are short-term dependencies in the data. Consider the case where the covariate vector is given by $z_1 = 2^{-1/2}, z_2 = -2^{-1/2}$, and $z_3, \ldots, z_n = 0$. Then, for the biased coin design,

$$z'\Sigma_T z = 2p \geq 1 + (2p - 1)^2 \tag{4.8}$$

(Problem 4.5). Hence, certain choices of z can behave far worse than Efron's solution ignoring short-range dependencies would suggest.

Markaryan and Rosenberger (2010) derive the exact variance–covariance matrix Σ_T and show that one of its eigenvalues is $2p$, which affirms that the maximal eigenvalue always exceeds $1 + (2p - 1)^2$. After extensive numerical evidence, they conclude that $2p$ is, indeed, the maximum eigenvalue.

Smith performs a formal spectral analysis of the generalized biased coin design, given in equation (3.25), assuming that $z_1, \ldots, z_n$ form a weakly stationary process. He concludes that the vulnerability to accidental bias is of the order

$$1 + \rho(1 + \rho)(1 + 2\rho)^{-1}n^{-1}\ln n + O(n^{-1}). \tag{4.9}$$

When $\rho = 1$, we have Wei's urn design, and equation (4.9) reduces to

$$1 + \frac{2}{3}\frac{\ln n}{n} + O(n^{-1}).$$

Efron's model for accidental bias describes a worst-case scenario in which an underlying covariate biases the treatment effect. The minimax solution minimizes the maximum effect of the covariate. This is very different from measuring the probability of a covariate imbalance, as we did in Section 4.3. It seems clear that, with the exception of the truncated binomial design, accidental bias is negligible for the restricted randomization procedures we explored. However, for covariates with drift or autocorrelation, our simulations show that covariate imbalances can be mitigated by using certain designs, such as the random block design or the accelerated biased coin design. Because these results are so similar to the balancing property on treatments, it appears that the balancing property of restricted randomization with respect to treatment assignments is strongly related to the balancing property on covariates.

4.7 CHRONOLOGICAL BIAS

Chronological bias was coined by Matts and McHugh (1978) to describe systemic temporal changes in the sequential accrual of subjects in clinical trials. They use as their example the maturation or deterioration of subjects, especially in situations where the recruitment period is long or the time between sequential entries in the trial is lengthy. We can think of chronological bias as a special case of accidental bias, where there is a time trend, or drift, in patient characteristics that may be related to the patient's outcome. In Section 4.3, we simulated the impact of a linear drift on a covariate imbalance. Under Efron's regression modeling approach, z could be represented as a time trend, and its impact on the treatment effect assessed.

Some preliminary simulation-based analyses have been presented in the literature addressing the impact of time trends for various restricted randomization procedures under a regression model approach. Rosenkranz (2011) investigated the impact of a linear time trend on the type I error rate of the standard t-test for complete randomization, the random allocation rule, truncated binomial design, Efron's biased coin design, and Wei's UD(0, 1). He concludes that the usual t-test attains the nominal size only for complete randomization and the random allocation rule, primarily due to the incorrect variance in the t-test when based on designs that do not have equiprobable sequences. He finds that the biased coin design and UD(0, 1) produce very conservative tests, and the truncated binomial design yields inflated type I error rate. Tamm and Hilgers (2014) investigate the permuted block design under linear, logarithmic, and stepwise time trends. They find that the t-test is very conservative, even for small time trends, unless the block size is very small. They suggest adjusting for the blocks in the analysis by adding block as a covariate in an analysis of variance.

The impact of time trends in biasing the treatment effect thus appears to be highly dependent on the particular randomization procedure employed. Determining theoretically which procedures mitigate the impact of time trends most is, therefore, an interesting open topic in both theory and practice.

4.8 PROBLEMS

4.1 Show that, under complete randomization, The Chebyshev bound holds in equation (4.5) if $Z_1, \dots, Z_n$ are independent. $N(\mu_i, \sigma_i^2)$, provided $\sum_{i=1}^n \sigma_i^2$ is $o(n^2)$.

4.2 Graph the simulated values of $P(|\bar{Z}_A - \bar{Z}_B| > 0.4)$ for $n = 50$ and 100 for the following restricted randomization procedures:

a. The random block design across values of $B_{max} = 1, 2, \dots, 10$;
b. Efron's biased coin design across values of $p = 0.55, 0.60, \dots, 0.95$.
c. Draw conclusions.

4.3 Show that 1 is the smallest possible value for the maximum eigenvalue of Σ_T when $T_j = 1$ or -1.

4.4
a. Show that the variance–covariance matrix for treatment allocation in permuted block randomization using the random allocation rule (within a block i of size $m = n/M$) is given by a block diagonal matrix with diagonal elements

$$\Sigma_{T,i} = \left(1 + \frac{1}{m-1}\right)I - \frac{1}{m-1}J, i = 1, \dots, M,$$

where I is the identity matrix and J is a matrix of 1's.
b. Show that the maximum eigenvalue of Σ_T is given by

$$\lambda = 1 + \frac{1}{m-1}.$$

c. Graph the vulnerability to accidental bias versus values of m.

4.5 Prove equation (4.8).

4.9 REFERENCES

BERRY, D. A. (1989). Comment: ethics and ECMO. *Statistical Science* **4** 306–310.
EFRON, B. (1971). Forcing a sequential experiment to be balanced. *Biometrika* **58** 403–417.
MARKARYAN, T. and ROSENBERGER, W. F. (2010). Exact properties of Efron's biased coin randomization procedure. *Annals of Statistics* **38** 1546–1567.
MATTS, J. P. and McHUGH, R. B. (1978). Analysis of accrual randomized clinical trials with balanced groups in strata. *Journal of Chronic Diseases* **31** 725–740.
RAO, C. R. (1973). *Linear Statistical Inference*. John Wiley & Sons, Inc., New York.
ROSENKRANZ, G. K. (2011). The impact of randomization on the analysis of clinical trials. *Statistics in Medicine* **30** 3475–3487.
ROSENBERGER, W. F. and RUKHIN, A. L. (2002). Bias properties and nonparametric inference for truncated binomial randomization. *Journal of Nonparametric Statistics* **15** 455–465.

SMITH, R. L. (1984). Sequential treatment allocation using biased coin designs. *Journal of the Royal Statistical Society B* **46** 519–543.

STEELE, J. M. (1980). Efron's conjecture on vulnerability to bias in a method for balancing sequential trials. *Biometrika* **67** 503–504.

TAMM, M. and HILGERS, R.-D. (2014). Chronological bias in randomized clinical trials arising from different types of unobserved time trends. *Methods of Information in Medicine*, **53**, 501–510.

ŽNIDARIČ, M. (2009). Asymptotic expansion for inverse moments of the binomial and Poisson distributions. *Open Statistics and Probability Journal* **1** 7–10.

4.10 APPENDIX

Here we present the SAS code to simulate the probability of a covariate imbalance when the random block design with parameter B_{max} is used. The user specifies B_{max} and the three seeds for the random number generation. The results are given in Table 4.1

```
data one;
 c=100000;
 Bmax=;
 epsilon=0.4;
 seed1=; seed2=; seed3=;
 array count (3) count1-count3;
 array sa (3) sa1-sa3;
 array sb (3) sb1-sb3;
 array z (3) z1-z3;
 array zbara (3) zbara1-zbara3;
 array zbarb (3) zbarb1-zbarb3;
 array diff (3) diff1-diff3;
 array prob (3) prob1-prob3;
 do n=50,100;
  do l=1 to 3;
   count(l)=0;
  end;
  do i=1 to c;
   na=0; nb=0;
   do l=1 to 3;
    sa(l)=0; sb(l)=0;
   end;
   cumblksize=0;
   do until (cumblksize > n);
    y=ranuni(seed1);
    do b=1 to Bmax;
     if (b-1)/Bmax < y < b/Bmax then blocksize=2*b;
    end;
    nablock=0; m=1;
    do j=cumblksize+1 to min(cumblksize+blocksize,n);
     z1=rannor(seed2);
     z2=rannor(seed2)+(4*j/n)-2;
     if j=1 then c0=rannor(seed2);
```

```
        c1=rannor(seed2);
        z3=c0+c1;
        c0=c1;
        x=ranuni(seed3);
        p=((blocksize/2)-nablock)/(blocksize-(m-1));
        if x < p then do;
         na+1; nablock+1; m+1;
         do l=1 to 3;
          sa(l)=sa(l)+z(l);
         end;
        end;
        else do;
         nb+1; m+1;
         do l=1 to 3;
          sb(l)=sb(l)+z(l);
          end;
         end;
        end;
        cumblksize=cumblksize+blocksize;
       end;
       do l=1 to 3;
        zbara(l)=sa(l)/na;
       zbarb(l)=sb(l)/nb;
       diff(l)=abs(zbara(l)-zbarb(l));
       if diff(l)>epsilon then count(l)+1;
       end;
      end;
      do l=1 to 3;
       prob(l)=count(l)/c;
      end;
      output;
     end;
proc print;
 var prob1 prob2 prob3;
 by n;
run;
```

5

Selection Bias

5.1 INTRODUCTION

Selection bias refers to biases that are introduced into an unmasked study because an investigator may be able to guess the treatment assignment of future patients based on knowing the treatments assigned to the past patients. Patients usually enter a trial sequentially over time. Staggered entry allows the possibility for a study investigator to alter the composition of the groups by attempting to guess which treatment will be assigned next. Based on whichever treatment is guessed to be assigned next, the investigator can then choose the next patient scheduled for randomization to be one whom the investigator considers to be better suited for that treatment. One of the principal concerns in an unmasked study is that a study investigator might attempt to "beat the randomization" and recruit patients in a manner such that each patient is assigned to whichever treatment group the investigator feels is best suited to that individual patient.

This type of guessing in an unmasked trial could introduce a bias in the composition of the treatment groups, which in turn could bias the study results. In principle, randomization is employed to mitigate this bias, and, indeed, this is a compelling reason to have randomized clinical trials. But the investigator may still be able to guess future treatment assignments with high probability, depending on the randomization procedure employed. As noted in Chapter 3, selection bias arises most frequently in the context of permuted block designs with fixed block sizes, as some treatment assignments will necessarily be deterministic to ensure balance within each block. The unmasking of the sequence of past treatment assignments can allow accurate prediction of future treatment assignments in the same block.

Classic scenarios for selection bias are part of the clinical trialist's folklore. An investigator for a pharmaceutical company, very anxious to see the company's latest pharmaceutical product succeed, guesses the randomization sequence and randomizes patients he or she deems more likely to respond positively to the new therapy

Randomization in Clinical Trials: Theory and Practice, Second Edition.
William F. Rosenberger and John M. Lachin.
© 2016 John Wiley & Sons, Inc. Published 2016 by John Wiley & Sons, Inc.

when he believes the new therapy to be next in the sequence. A sympathetic nurse coordinator tries to assign a favorite patient to the new therapy rather than placebo. These scenarios are likely to be more the result of subconscious preferences than deliberate dishonesty. Also, any deliberate guessing is likely to be inaccurate, as randomization procedures can be complicated and the investigator may not understand fully the subtleties of the particular procedure used. However, as Smith (1984a) points out, they do not have to be right all the time: it is sufficient that they make more right guesses than wrong guesses. While multicenter trials may make it difficult to determine what is going on in other centers, stratification within clinical center eliminates this protection.

Berger (2005a) distinguishes among three orders of selection bias. First-order selection bias interferes with the internal validity of the trial when patient selection precedes randomization. Second-order selection bias occurs when advance randomization is used in a trial without allocation concealment. Third-order selection bias occurs with imperfect or unsuccessful allocation concealment or in unmasked trials. In this chapter, we will be concerned with third-order selection bias, as we assume that some form of allocation concealment is used (even in an unmasked trial). Senn (2005) describes third-order selection bias in the context of clinical trials that are "randomized by intention but not execution."

The great clinical trialist Chalmers (1990) was convinced that the elimination of selection bias is the most essential requirement for a good clinical trial. He was especially concerned that there are too many loopholes in eligibility criteria and in the rejection of patients during the screening phase, during which the physician could project his or her doubts to the patients while seeking consent. Even when the randomization sequences are intended to be masked, it is not unusual for patients to be unmasked during the course of the trial, due to either adverse events known to be highly associated with one of the treatments, or life-threatening emergencies requiring unmasking, or distinguishing features of the masked treatment, such as taste. Regardless of how it arises, selection bias can result in covariate imbalances and inflated type I error rates (Proschan, 1994; Berger and Exner, 1999). It is closely related to the previous chapter on balancing with respect to covariates; as Berger (2005a) states, randomization is necessary to ensure that baseline imbalances are random, but not sufficient due to selection bias.

In this chapter, we examine a simple model for susceptibility to selection bias, developed by Blackwell and Hodges (1957), for randomization procedures that are intended to promote balance. The model assumes that the investigator will guess the randomization sequence and attempt to put each patient on the treatment that he or she believes is better for that patient. We then show that the Blackwell–Hodges model can be interpreted as a measure of predictability of a randomization procedure that does not involve a guessing strategy at all. The measure compares the conditional allocation probabilities to the unconditional allocation probabilities. We compare various restricted randomization procedures with respect to this measure. Finally, we describe tests for selection bias.

5.2 THE BLACKWELL–HODGES MODEL

The Blackwell–Hodges model for selection bias assumes that random treatment assignment is independent of the patient characteristics and responses, meaning that an adaptive procedure is not employed. Suppose the primary outcome of a trial is a random variable Y, and the null hypothesis is true; that is, $E(Y|A) = E(Y|B) = \mu$. The experimenter wishes to bias the study by selecting a patient with a higher value of $E(Y)$ when he guesses that A is the next treatment, designated as the guess a, and a lower value of $E(Y)$ when he guesses that B is the next treatment, designated as the guess b. Let $E(Y|a) = \mu + \Delta$ be the expected value of the response when the experimenter guesses a, and let $E(Y|b) = \mu - \Delta$ be the expected value of the response when the experimenter guesses b. Of course, the experimenter's guess may be wrong. The accuracy of guessing is described by the parameters (α, β), which represent the probabilities of correct guesses of treatment A and B, respectively. For illustration, assume that n is even, and the number assigned to each treatment is fixed at $n/2$ (the model will also apply when the limiting proportions are $1/2$). Then the expected numbers of guesses, and the expected values of each, are represented in Figure 5.1.

Let G be the total number of correct guesses, and let $\bar{Y}_A$ and $\bar{Y}_B$ be the treatment group means among those randomized to A and B, respectively. Then from Figure 5.1, the expected number of correct guesses is $E(G) = (\alpha + \beta)n/2$. The possible bias introduced by correct guesses is represented by the expected treatment group means among those randomized to each treatment. These are

$$E(\bar{Y}_A) = \frac{(\alpha n/2)(\mu + \Delta) + \left[(1 - \alpha)(n/2)\right](\mu - \Delta)}{n/2} = 2\alpha\Delta + (\mu - \Delta),$$

$$E(\bar{Y}_B) = \frac{\left[(1 - \beta)n/2\right](\mu + \Delta) + \left[\beta n/2\right](\mu - \Delta)}{n/2} = -2\beta\Delta + (\mu + \Delta).$$

Then the expected treatment difference is given by

$$E(\bar{Y}_A - \bar{Y}_B) = 2\Delta(\alpha + \beta - 1) = 2\Delta\frac{E\left(G - \frac{n}{2}\right)}{\frac{n}{2}}. \tag{5.1}$$

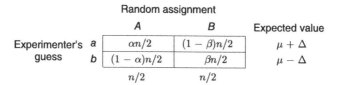

	Random assignment		
	A	**B**	Expected value
Experimenter's **a**	$\alpha n/2$	$(1 - \beta)n/2$	$\mu + \Delta$
guess **b**	$(1 - \alpha)n/2$	$\beta n/2$	$\mu - \Delta$
	$n/2$	$n/2$	

Fig. 5.1 The Blackwell–Hodges model for selection bias.

In (5.1), the investigator's bias 2Δ is the quantity introduced by attempts to beat the randomization. If $\Delta = 0$, we have no bias, and hence, the investigator cannot bias the study since there is truly no differential treatment effect between the groups guessed to be assigned to A and the groups guessed to be assigned to B. The *expected bias factor* is the remaining term

$$E(F) = E\left(G - \frac{n}{2}\right). \tag{5.2}$$

If the experimenter guesses completely at random, then $\alpha = \beta = 1/2$ and $E(F) = 0$. These results also apply to unbalanced randomization (Problem 6.1).

Blackwell and Hodges (1957) show that, under a restricted randomization procedure with balanced allocation, the optimal strategy for the experimenter upon randomizing the jth patient is to guess treatment A when $N_A(j - 1) < N_B(j - 1)$ and guess treatment B when $N_A(j - 1) > N_B(j - 1)$. When there is a tie, the experimenter guesses with equal probability. Blackwell and Hodges call this the *convergence strategy*.

We see from (5.1) that the expected bias factor $E(F)$ for a randomization procedure where the experimenter employs the convergence strategy can be obtained as follows. For the jth patient, the experimenter guesses treatment A when $N_A(j - 1) < N_B(j - 1)$ and guess treatment B when $N_A(j - 1) > N_B(j - 1)$. Among these guesses, now call a correct guess a *hit* and an incorrect guess a *miss*. If $N_A(j - 1) = N_B(j - 1)$ (i.e., we have a *tie*), then the investigator has no basis for a guess and arbitrarily chooses either A or B systematically. Let H, M, and T denote the total number of hits, misses, and ties, in an n-patient randomization stream, respectively. For ties, by chance, $T/2$ are expected to be guessed correctly. Therefore, from the aforementioned,

$$E(G) = \frac{(\alpha + \beta)n}{2} = E\left[H + \frac{T}{2}\right].$$

Since $n = E(H + M + T)$, we have

$$E(F) = E(G) - \frac{n}{2} = \frac{E(H - M)}{2}. \tag{5.3}$$

Equation (5.3) allows us to assess the expected bias factor for any given sequence of random assignments.

Under this model, for any double-masked randomization, regardless of the method of treatment assignment, since N_A and N_B are unknown to the investigator, the expected number of correct guesses $E(G)$ is simply $n/2$, in which case $E(F) = 0$. The issue, however, is the potential for selection bias in an unmasked study. With complete randomization, $E(F) = 0$ because there is a fixed probability of $1/2$ of assignment to A for all allocations, and thus, future random assignments are not in any way predictable based on past assignments. Thus, complete randomization eliminates the expected potential for selection bias. However, this is not the case with restricted randomization procedures, which are designed to eliminate or reduce the probability of treatment imbalances. Such sequences are to some degree predictable and, thus, are subject to selection bias. In the following sections, the precise expressions for $E(F)$ are presented for various restricted randomization designs.

Stigler (1969) (see also Wei, 1978a) describes the Blackwell–Hodges model in terms of a minimax strategy: they wish to find a design that minimizes the maximum risk, as given by $E(F)$. Stigler proposes that, rather than bias the experiment by Δ on each trial, assume that the investigator picks a subject with expected response between $\mu - \Delta$ and $\mu + \Delta$. Thus, the investigator may choose not to bias the experiment at all (selecting a subject with expected response μ), or, at worst, choose according to the Blackwell–Hodges model. Stigler's justification for this more conservative model is that investigators will tend to be timid in realistic situations. He offers the *proportional convergence strategy* to express this timidity. If the investigator observes, after i trials, j treatment A assignments and $i - j$ treatment B assignments, he will then select a subject with expected response

$$\mu + 2\Delta \left(\frac{\frac{n}{2} - j}{n - i} - \frac{1}{2} \right). \tag{5.4}$$

5.3   PREDICTABILITY OF A RANDOMIZATION SEQUENCE

Selection bias occurs as a result of the predictability of a randomization sequence in conjunction with the selection of patients by the investigator. In complete randomization, each sequence is completely unpredictable. The degree of predictability of a sequence can be measured in the absence of a guessing strategy. Berger, Ivanova, and Knoll (2003) suggest that predictability can be measured in restricted randomization by the difference between the conditional probability of a treatment assignment and its unconditional probability $(1/2)$. One measure of the predictability of a sequence is given by

$$\rho_{\text{PRED}} = \sum_{j=1}^{n} E \left| \phi_j - \frac{1}{2} \right|, \tag{5.5}$$

which is 0 for complete randomization and takes its maximum value, $n/2$ if the sequence is deterministic. This is similar to a metric proposed by Chen (2000). (Other authors have found the expected proportion of assignments that are predictable with probability one; see Dupin-Spriet, Fermanian, and Spriet (2004).) However, it turns out that ρ_{PRED} in (5.5) is mathematically equivalent to the expected bias factor in (5.3) for restricted randomization procedures, as we now show.

For any restricted randomization procedure defined by $\phi_j, j = 1, \ldots, n$ in Chapter 3, we have the following equivalence. Let $I(\cdot)$ be the indicator function. The expected number of hits and misses, given $\phi_j, j = 1, \ldots, n$ is computed as

$$E(H|\phi_1, \ldots, \phi_n) = \sum_{j=1}^{n} \left[\frac{1}{2} I \left(\phi_j = \frac{1}{2} \right) + \phi_j I \left(\phi_j > \frac{1}{2} \right) + (1 - \phi_j) I \left(\phi_j < \frac{1}{2} \right) \right]$$

and

$$E(M|\phi_1, \ldots, \phi_n) = \sum_{j=1}^{n} \left[\frac{1}{2} I \left(\phi_j = \frac{1}{2} \right) + (1 - \phi_j) I \left(\phi_j > \frac{1}{2} \right) + \phi_j I \left(\phi_j < \frac{1}{2} \right) \right].$$

The expected selection bias factor in (5.3) is then calculated as

$$E(F) = EE(F|\phi_1, \ldots, \phi_n) = E \left(\frac{E(H|\phi_1, \ldots, \phi_n) - E(M|\phi_1, \ldots, \phi_n)}{2} \right)$$

$$= E \sum_{j=1}^{n} \left[\left(\phi_j - \frac{1}{2} \right) I \left(\phi_j > \frac{1}{2} \right) + \left(\frac{1}{2} - \phi_j \right) I \left(\phi_j < \frac{1}{2} \right) \right]$$

$$= \sum_{j=1}^{n} E \left| \phi_j - 1/2 \right|$$

$$= \rho_{PRED}.$$

5.4 SELECTION BIAS FOR THE RANDOM ALLOCATION RULE AND TRUNCATED BINOMIAL DESIGN

In this section, we compare the random allocation rule (RAR) and the truncated binomial design (TBD) with respect to the expected selection bias factor. It is a somewhat surprising result that the TBD has a smaller expected bias factor than the RAR.

We can show that, for the RAR, when the convergence strategy is employed,

$$E(F) = \frac{2^{n-1}}{\binom{n}{n/2}} - \frac{1}{2}, \tag{5.6}$$

as follows.

Think of the RAR as a random walk on a plane starting at point $(0, 0)$ and terminating at $(n/2, n/2)$, moving one unit to the right when A is chosen and one unit up when B is chosen. When the walk hits the diagonal, $N_A = N_B$, and the experimenter guesses at random. Away from the diagonal, when the convergence strategy is employed, the experimenter always guesses that the walk moves toward the diagonal. Since the walk begins and ends on the diagonal, it follows that the walk moves toward the diagonal exactly $n/2$ times. Consequently, the experimenter is right at least $n/2$ times. In addition, the experimenter is right, on average, half the time the walk is on the diagonal. Let T be the number of ties. Then

$$E(G) = \frac{n}{2} + \frac{E(T)}{2}.$$

It remains to find $E(T)$. The distribution of T was given by Feller (1950), (but apparently not in later editions):

$$\Pr(T = t) = \frac{2^t \left[\binom{n-t-2}{n/2-t} - \binom{n-t-2}{n/2-t-2} \right]}{\binom{n}{n/2}}. \tag{5.7}$$

Using (5.7), one can derive

$$E(T) = \frac{2^n}{\binom{n}{n/2}} - 1, \tag{5.8}$$

(Problem 6.2), and (5.6) follows immediately.

The TBD of Chapter 3 was proposed by Blackwell and Hodges as an alternative to the RAR that would provide less susceptibility to selection bias. For this design, the optimal strategy is to guess the same treatment until $n/2$ of one treatment have been assigned. Then the experimenter should switch to the treatment arm that has less than $n/2$ since all future assignments are known with certainty. We can use the distribution of the number of deterministic selections in the tail, X, given (3.6), to determine $E(F)$. Since the total number of correct guesses under the TBD is $G = (n - X)/2 + X$, we therefore have

$$E(G) = \frac{n}{2} + \frac{E(X)}{2}. \tag{5.9}$$

Substituting equation (3.7) into (5.9) gives the result

$$E(F) = \frac{n}{2^{n+1}} \binom{n}{n/2}. \tag{5.10}$$

It turns out that, while the results of Chapter 4 show that the TBD has a high degree of accidental bias, it always has a smaller value of $E(F)$ than the RAR, which we now demonstrate (although the RAR does better than the TBD when used under Stigler's (1969) proportional convergence strategy in (5.4)). By (5.6) and (5.10), we must show that

$$\frac{n}{2^{n+1}} \binom{n}{n/2} \le \frac{2^{n-1}}{\binom{n}{n/2}} - \frac{1}{2}$$

or, equivalently,

$$n \binom{n}{n/2}^2 + 2^n \binom{n}{n/2} - 2^{2n} \le 0, \tag{5.11}$$

for $n \geq 0$, even. For $n = 2$, one can see that the inequality is sharp. Assume that (5.11) holds. Then we must show

$$(n + 2)\binom{n + 2}{(n + 2)/2}^2 + 2^{n+2}\binom{n + 2}{(n + 2)/2} - 2^{2n+4} \leq 0. \tag{5.12}$$

Noting that

$$\binom{n + 2}{(n + 2)/2} = 4\left(\frac{n + 1}{n + 2}\right)\binom{n}{n/2},$$

the left-hand side of (5.12) is equal to

$$16\frac{(n + 1)^2}{n + 2}\binom{n}{n/2}^2 + 16\left(\frac{n + 1}{n + 2}\right)2^n\binom{n}{n/2} - 16 \times 2^{2n}$$

$$= 16\left[n\binom{n}{n/2}^2 + 2^n\binom{n}{n/2} - 2^{2n}\right] + \frac{16}{n + 2}\binom{n}{n/2}\left[\binom{n}{n/2} - 2^n\right].$$

The first term is ≤ 0 by (5.11). For the second term,

$$\binom{n}{n/2} \leq 2^n$$

by a similar induction argument.

Figure 5.2 plots the expected bias factor for the RAR (5.6) and the TBD (5.10).

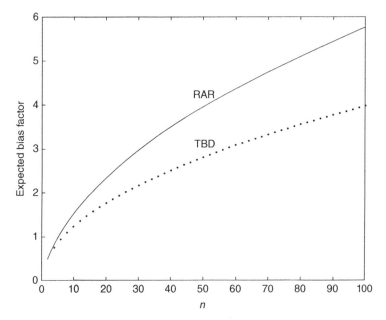

Fig. 5.2 *Expected bias factor for the random allocation rule (RAR) and the truncated binomial design (TBD) across values of n.*

5.5   SELECTION BIAS IN A PERMUTED BLOCK DESIGN

It also follows from the aforementioned result that the permuted block design will have less potential for selection bias when allocations are made using a truncated binomial than when using a RAR. The explicit relationships follow.

5.5.1   Permuted blocks using the random allocation rule

Under the original Blackwell–Hodges model, each of the M blocks in a permuted block design has a potential selection bias equal to that of a RAR of the same size. Thus, from (5.6), the expected bias factor, for a permuted block design with M blocks of equal size $m = n/M$, is

$$E(F) = M \left(\frac{2^{m-1}}{\binom{m}{m/2}} - \frac{1}{2} \right). \tag{5.13}$$

Comparing (5.13) to (5.6), it follows that the expected bias factor for a permuted block randomization with $M > 1$ is always greater than that for the RAR (where $M = 1$). For example, if $n = 100$, then for a RAR $E(F) = 5.78$, whereas for a permuted block design with five blocks of size 20, $E(F) = 11.69$.

   Figure 5.3 shows the total expected bias factor for increasing n for various block sizes of $m = 2, 4, 6, 8,$ or 10.

5.5.2   Permuted blocks with truncated binomial randomization

An alternative strategy to lessen the susceptibility to selection bias of a permuted block design is to generate the assignments within each block using a truncated binomial design. For a permuted block design with block size m, the expected selection bias due to predictions, which can be made with certainty under truncated binomial sampling, is computed from (5.10), which yields

$$E(F) = M \left[\frac{m}{2^{m+1}} \binom{m}{m/2} \right].$$

Figure 5.4 compares the expected bias factor across fixed block sizes and $n = 100$, using the RAR and the truncated binomial rule. One can see that the truncated binomial rule results in smaller expected bias than the RAR when used in permuted blocks.

5.5.3   Random block design

One strategy that has been widely used in an effort to reduce the potential for selection bias with the permuted block design is to employ a random block design with random block sizes. Unfortunately, this strategy still yields a substantial potential for selection

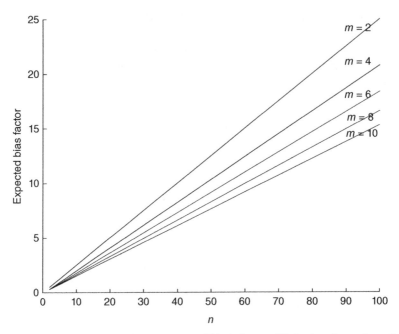

Fig. 5.3   *Expected bias factor for the permuted block design (filled using the random alloca-tion rule) with block sizes m = 2, 4, 6, 8, 10 across values of n.*

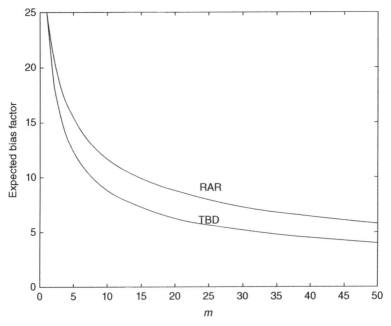

Fig. 5.4   *The effect of block size on the expected bias factor for the permuted block design with n = 100, comparing the random allocation rule (RAR) and the truncated binomial design (TBD).*

bias that is not significantly better than that of a permuted block design at a fixed block size equal to the average block size in the random block design.

In general, for a permuted block design with possibly unequal block sizes $m_i, i = 1, 2, \ldots, M$, the overall expected selection bias factor is the sum of the individual block selection bias factors:

$$E(F) = \sum_{i=1}^{M} \left(\frac{2^{m_i-1}}{\binom{m_i}{m_i/2}} - \frac{1}{2} \right),$$

and similarly if blocks are filled using the TBD. This formula does not apply to the random block design. It is not possible to write the form of $E(F)$ using a simple sum, since blocks are selected randomly, and one must average over all possible combinations of block sizes and include a random unfilled final block to achieve n patients. If we ignore the final incomplete block, and just stop the trial after a random number $J - 1$ blocks of size $M_i, i = 2, 4, \ldots, 2B_{max}$, where $J = \min\{j : \sum_{i=1}^{j} M_i > n\}$, then the expected bias factor can be written as

$$E(F) = E_{M_i,J} \sum_{i=1}^{J-1} \left(\frac{2^{M_i-1}}{\binom{M_i}{M_i/2}} - \frac{1}{2} \right).$$

While such a procedure could never be used in an actual clinical trial, the expression allows us to see that the use of random block sizes does not eliminate the potential for selection bias. There is still an additive component over the blocks, and these components are averaged over the block sizes and number of blocks. One can presume that the expected bias factor will be similar to that of the permuted block design where the block size is roughly equivalent to the average block size of the random block design, B_{max}.

Shao (2015) finds the expected bias factor using equations (5.5) and (3.15), successively conditioning on the block size (B_j) and position number within the block (R_j) of the jth patient:

$$\rho_{PRED} = \sum_{j=1}^{n} E \left| \phi_j - \frac{1}{2} \right| = \sum_{j=1}^{n} E_{B_j} E_{R_j|B_j} E \left\{ \left| \phi_j - \frac{1}{2} \right| | R_j, B_j \right\}.$$

The result, in closed form, would take up considerable space.

5.5.4 Conclusions

When using a permuted block design, selection bias can be effectively reduced by employing the truncated binomial randomization procedure within blocks rather than the RAR. However, as pointed out in Chapter 4, the risk of accidental bias is much higher for truncated binomial randomization. Selecting block sizes randomly does not eliminate selection bias or reduce it substantially.

5.6 SELECTION BIAS FOR OTHER RESTRICTED RANDOMIZATION PROCEDURES

5.6.1 Efron's biased coin design

Markaryan and Rosenberger derive an exact expression for $E(F)$ under Efron's biased coin design, using the convergence strategy, and tabulate the results. One can obtain the result by using the expression in (5.5):

$$\rho_{\text{PRED}} := \left(p - \frac{1}{2}\right) \sum_{j=1}^{n} (1 - P(D_{j-1} = 0)),$$

where $P(D_n = 0)$ can be found in its exact form in Theorem 1 of Markaryan and Rosenberger (2010). No exact expression is known for generalizations of Efron's biased coin design, and this should be considered an open problem.

5.6.2 Wei's urn design

For the urn randomization, $\text{UD}(\alpha, \beta)$, the probability of assignment to treatment A fluctuates around $1/2$ to a degree proportional to the extent of imbalance. Since $N_A(n)/n \to 1/2$ as $n \to \infty$, the degree of fluctuation around $1/2$ decreases as n increases. Thus, future assignments are more predictable early in the sequence of assignments, and the predictability of assignments decreases as n increases. In turn, the potential for selection bias increases initially but then converges to an asymptote as n increases.

Unlike for the biased coin design, a closed-form expression for the expected bias factor has not been previously derived. However, Wei (1977) derived a recursive formula. The imbalance after n assignments is $D_n = N_A(n) - N_B(n)$. For any value $|D_n| = d$, $0 \le d \le n$, the probability of a correct guess on the $(n+1)$th allocation, conditional on $|D_n| = d$, is denoted as

$$P(C_{n+1}|d) = \frac{1}{2} + \frac{\beta d}{2(2\alpha + \beta n)}.$$

Therefore, unconditionally, the probability of a correct guess on the $(n+1)$th assignment is denoted

$$P(C_{n+1}) = \sum_{d=0}^{n} P(C_{n+1}|d) \Pr(|D_n| = d) = \frac{1}{2} + \frac{\beta E|D_n|}{2(2\alpha + \beta n)},$$

where $E(|D_n|) = \sum_{d=0}^{n} d \Pr(|D_n| = d)$ can be obtained by the recursive relationship in (3.22). Therefore, the expected bias factor after n assignments for

$$E(F) = \sum_{i=1}^{n} P(C_i) - \frac{n}{2}. \tag{5.14}$$

5.6.3 Smith's design

Smith (1984a) considers the limiting value of the selection bias for the class of generalized biased coin designs, given in (3.24). His measure of selection bias is given by

$$\lim_{n \to \infty} \frac{2\Delta E(F)}{n}. \tag{5.15}$$

Under this definition, the RAR and TBDs each have expected selection bias of order $O(n^{-1/2})$ (Problem 6.3).

Using the notation in Section 3.8, the probability of a correct guess for the $(j + 1)$th patient, using the convergence criterion, is given by

$$\max\left\{ p\left(\frac{N_A(j) - N_B(j)}{j} \right), 1 - p\left(\frac{N_A(j) - N_B(j)}{j} \right) \right\},$$

and therefore the expected number of correct guesses minus the expected number of incorrect guesses in n patients is

$$\sum_{j=1}^{n-1} \left| 2p\left(\frac{N_A(j) - N_B(j)}{j} \right) - 1 \right| = 2E\left(G - \frac{n}{2} \right).$$

Then the expected selection bias is

$$n^{-1} \Delta \sum_{j=1}^{n-1} \left| 2p\left(\frac{N_A(j) - N_B(j)}{j} \right) - 1 \right|. \tag{5.16}$$

Smith (1984b) shows that (5.16) is approximately

$$\Delta \rho \left(\frac{2}{n\pi(1 + 2\rho)} \right)^{1/2}, \tag{5.17}$$

as $n \to \infty$, where ρ is defined in (3.25). As with the RAR and the TBD, this result is $O(n^{-1/2})$. Using this result, we can directly compare generalized biased coin designs for various values of ρ. For example, when $\rho = 2$, we have approximately 1.55 times the bias of Wei's urn design with $\alpha = 0$ ($\rho = 1$).

5.7 SIMULATION RESULTS

In Table 5.1, we simulate the expected bias factor for different restricted randomization procedures for $n = 50, 100$ and $100,000$ replications. The procedures compared include complete randomization, Wei's UD(0, 1), Smith's design with $\rho = 5$, Efron's biased coin design ($p = 2/3$), the big stick design and biased coin design with imbalance intolerance (both with $b = 3$), the accelerated biased coin design with $a = 10$,

Table 5.1 *Simulated expected selection bias factor, n = 50, 100, based on* 100, 000 *replications.*

Randomization Procedure	$n = 50$	$n = 100$
Complete	0.10	1.47
UD(0, 1)	3.00	4.36
Smith ($\rho = 5$)	6.54	10.01
BCD(2/3)	6.09	12.34
Big Stick ($b = 3$)	3.95	8.11
BCDII(2/3, 3)	7.00	14.12
ABCD(10)	6.00	12.24
Hadamard	6.77	13.68
Maximal procedure ($b = 3$)	6.61	12.86
Permuted block design ($m = 2$, RAR)	12.50	25.00
Permuted block design ($m = 4$, RAR)	10.16	20.83
Permuted block design ($m = 4$, TBD)	8.99	18.75
Permuted block design ($m = 8$, RAR)	8.04	16.22
Permuted block design ($m = 8$, TBD)	6.56	13.13
Random block design ($B_{max} = 3$, RAR)	10.01	20.16
Random block design ($B_{max} = 3$, TBD)	8.94	18.06
Random block design ($B_{max} = 10$, RAR)	6.71	13.75
Random block design ($B_{max} = 10$, TBD)	5.16	10.77

the maximal procedure with $b = 3$, Hadamard randomization using the matrix in (3.13), the permuted block design with $m = 2, 4, 8$ filled with both the RAR and the TBD, and the random block design with $B_{max} = 3, 10$, filled using the RAR and TBD. An example of the SAS code to run this simulation is given for the UD(0, 1) in the appendix in Section 5.11.

If avoiding selection bias is the most important consideration in a particular clinical trial, clearly the UD(0, 1) is the best of all the procedures. Only the big stick design is close. As discussed earlier, the random block design does not perform better than the permuted block design with fixed block size equal to the average block size of the random blocks. The maximal procedure, initially proposed as an alternative to blocked randomization, but less vulnerable to selection bias, is about equivalent to a permuted block design, filled using the TBD, for $m = 8$, and slightly better than Hadamard randomization. The fact that the maximal procedure has a major logistical constraint that n must be known in advance, combined with its higher expected selection bias factor, hinders its usefulness in practice.

We note that the procedures that performed best with respect to balance in Table 3.4 performed the worst here and vice versa. This can easily be seen with the two extremes: complete randomization is the worst for balance, but has no vulnerability to selection bias. An alternating deterministic sequence has perfect balance but the worst possible vulnerability to selection bias. We will discuss this trade-off between criteria for "good" and "bad" randomization procedures in Chapter 8.

5.8   CONTROLLING AND TESTING FOR SELECTION BIAS IN PRACTICE

Berger and Exner (1999) describe several measures that can be taken in a randomized clinical trial to avoid selection bias. Among these are:

1. Maintain a registry of all screened patients, along with a unique identifier, date and time of screening, well-documented rationale for enrollment decisions, and baseline measurements. Having the date and time of screening allows one to determine the treatment group to which the patient would have been enrolled had he or she been randomized.
2. If permuted block randomization is used and a treatment code is unmasked for a patient before the completion of enrollment in that patient's block, redefine the block to consist of only those patients enrolled at the time of the unmasking and cease enrollment to the block. Proceed to the next block, possibly appending additional blocks to ensure adequate enrollment.
3. Consider excluding from enrollment decisions investigators who evaluate patients.
4. Do not reuse patient numbers for those patients who have dropped out, and do not bypass new randomizations by giving the same treatment to a replacement patient.

Berger and Exner (1999) also suggest testing for selection bias by examining, within each treatment group, the effect on the response variable of the probability that a patient receives the active treatment, adjusted for the treatment assigned. If $Y_1, \ldots, Y_n$ are normally distributed responses, then from the analysis of covariance model

$$E(Y_j) = \alpha T_j + \beta \phi_j,$$

we conduct a test of $\beta = 0$ as the test of unobservable selection bias. If $Y_1, \ldots, Y_n$ are binary or from some other distribution, then generalized linear models can be used similarly.

While this formulation applies to any restricted randomization procedure, the Berger–Exner test was proposed originally in the context of the permuted block design. In this context, they suggest that the test be used in conjunction with more common baseline comparisons to also detect observable selection bias. These baseline comparisons should be performed once using the subset of patients for whom $\phi_j \neq 1/2$, in order to detect selection bias arising from the subversion of allocation concealment within blocks, and then again among all patients, to detect selection bias arising from the direct subversion of allocation concealment. If selection bias is found, it may be of interest to do a treatment comparison only among patients for whom $\phi_j = 1/2$.

Ivanova, Barrier, and Berger (2005) present a likelihood ratio test that detects observable selection bias and tests the treatment effect, adjusted for selection bias. They use a logistic regression model under a randomized block design and assume a Blackwell–Hodges model for selection bias under binary responses. The simulated tests have a chi-square distribution. For normally distributed responses,

Kennes, Rosenberger and Hilgers (2015) proves that the tests are asymptotically chi-square and give closed-form statistics and confidence intervals.

5.9 PROBLEMS

5.1 Derive the expected bias factor from the Blackwell–Hodges model when there is fixed unbalanced allocation.

5.2 Derive equation (5.8) from (5.7) (Blackwell and Hodges, 1957).

5.3 Use Stirling's formula to show that $E(F)/n$ for the RAR and TBD is of order $O(n^{-1/2})$.

5.4 Plot the expected selection bias, $E(F)/n$, versus p for Efron's biased coin design. Is $p = 2/3$ a reasonable choice when considering just selection bias? What about when considering the trade-off between balance and selection bias?

5.5 Read Berger (2005b) and the corresponding discussion. Prepare a two-page position paper addressing the following issues:

a. Is selection bias a real problem in modern randomized clinical trials? Give evidence for and against, and draw conclusions.
b. Assuming that selection bias is a compelling problem, what are the best means to control it?

5.10 REFERENCES

BERGER V. W. (2005a). *Selection Bias and Covariate Imbalances in Randomized Clinical Trials.* John Wiley & Sons, Ltd: Chichester.

BERGER, V. W. (2005b). Quantifying the magnitude of baseline covariate imbalances resulting from selection bias in randomized clinical trials. *Biometrical Journal* **47** 119–139 (with discussion).

BERGER, V. W. AND EXNER, D. V. (1999). Detecting selection bias in randomized clinical trials. *Controlled Clinical Trials* **20** 319–327.

BERGER, V. W., IVANOVA, A., AND KNOLL, M. D. (2003). Minimizing predictability while retaining balance through the use of less restrictive randomization procedures. *Statistics in Medicine* **42** 3017–3028.

BLACKWELL, D. AND HODGES, J. L. (1957). Design for the control of selection bias. *Annals of Mathematical Statistics* **28** 449–460.

CHALMERS, T. C. (1990). Discussion of biostatistical collaboration in medical research by Jonas H. Ellenberg. *Biometrics* **46** 20–22.

CHEN, Y.-P. (2000). Which design is better? Ehrenfest urn versus biased coin. *Advances in Applied Probability* **32** 738–749.

DUPIN-SPRIET, T., FERMANIAN, J., AND SPRIET, A. (2004). Quantification of predictability in clinical trials using block randomization. *Drug Information Journal* **38** 127–133.

FELLER, W. (1950). *An Introduction to Probability Theory and Its Application*, Vol. **I**. John Wiley & Sons, Inc., New York.

IVANOVA, A., BARRIER, R. C., AND BERGER, V. W. (2005). Adjusting for observable selection bias in block randomized trials. *Statistics in Medicine* **24** 1537–1546.

KENNES, L. N., ROSENBERGER, W. F., AND HILGERS, R. D. (2015). Inference for blocked randomization under a selection bias model. Biometrics, in press.

MARKARYAN, T. AND ROSENBERGER, W. F. (2010). Exact properties of Efron's biased coin randomization procedure. *Annals of Statistics* **38** 1546–1567.

PROSCHAN, M. (1994). Influence of selection bias on type I error rate under random permuted block designs. *Statistica Sinica* **4** 219–231.

SENN, S. (2005). Comment. *Biometrical Journal* **47** 133–135.

SHAO, H. (2015). Randomness and Variability in Restricted Randomization. Fairfax: George Mason University (doctoral dissertation).

SMITH, R. L. (1984a). Sequential treatment allocation using biased coin designs. *Journal of the Royal Statistical Society B* **46** 519–543.

SMITH, R. L. (1984b). Properties of biased coin designs in sequential clinical trials. *Annals of Statistics* **12** 1018–1034.

STIGLER, S. M. (1969). The use of random allocation for the control of selection bias. *Biometrika* **56** 553–560.

WEI, L. J. (1977). A class of designs for sequential clinical trials. *Journal of the American Statistical Association* **72** 382–386.

WEI, L. J. (1978a). On the random allocation design for the control of selection bias in sequential experiments. *Biometrika* **65** 79–84.

5.11 APPENDIX

Here we present the SAS code to simulate the expected selection bias factor when the UD(0, 1) restricted randomization procedure is used. The user specifies B_{max} and the three seeds for the random number generation. Results are given in Table 5.1.

```
data one;
do n=50,100;
 do i=1 to 100000;
  nb=0;
  hits=0;
  misses=0;
  na=0;
  seed=;
  do j=1 to n;
   if na=nb then guess=0;
   if na > nb then guess=-1;
   if na < nb then guess=1;
   if j=1 then p=1/2;
   else p=nb/(j-1);
   x=ranuni(seed);
   if x > p then do;
    t=-1; nb+1;
   end;
   else do;
```

```
    t=1; na+1;
  end;
  if guess=t then hits+1;
  if (guess ne 0 and guess ne t) then misses+1;
 end;
 bias=(hits-misses)/2;
 output;
 end;
end;
proc means n mean;
 var bias;
 by n;
run;
```

6

Randomization as a Basis for Inference

6.1 INTRODUCTION

In Chapters 4 and 5, we described several types of bias that can result in biomedical studies and showed how randomization can mitigate these biases. The second major contribution of randomization is that it can be used as a basis for inference at the conclusion of the trial. Analyses based on a randomization model are completely different from traditional analyses using hypotheses tests of population parameters under the Neyman–Pearson paradigm. In this chapter, we will explore the differences between the randomization model and the population model. In so doing, we will develop the principles of randomization-based inference using randomization tests, originally proposed in the early part of the last century by Fisher (1935). A warning to the reader: the Fisher randomization test has contributed to much controversy in the statistical world over recent years. In fact, Fisher himself was somewhat contradictory in his later writings on the subject. For an entertaining and heated debate on the subject, the interested reader is referred to Basu (1980). It should be clear that the authors of this book support randomization-based inference, and the reader should be thus informed. We feel that randomization-based inference is a useful alternative to, or complement to, traditional population model-based methods.

6.2 THE POPULATION MODEL

The most commonly used basis for the development of a statistical test is the concept of a *population model*, where it assumed that the sample of patients is representative of a reference population and that the patient responses to treatment are independent and identically distributed from a distribution dependent on unknown population parameters. In the population model, n_A and n_B patients are randomly sampled

Randomization in Clinical Trials: Theory and Practice, Second Edition.
William F. Rosenberger and John M. Lachin.
© 2016 John Wiley & Sons, Inc. Published 2016 by John Wiley & Sons, Inc.

from an infinite population of patients on treatment A and treatment B, respectively. Then the $n_i, i = A, B$ patient responses $(Y_{i1}, \ldots , Y_{in_i})$ can be treated as independent and identically distributed according to some probability distribution $G(y|\theta_i)$ having parameter θ_i. The population model is shown on the left side in Figure 6.1. Under this assumed distribution, it is then a direct matter to construct hypothesis tests comparing the treatment effects, under the Neyman–Pearson lemma, such as

$$H_0 : \theta_A = \theta_B \text{ versus } H_A : \theta_A \neq \theta_B,$$

if θ_i is a scalar. It can also be vector-valued, such as the case where G is normally distributed and $\theta_i = (\mu_i, \sigma^2)$. For this example, the t-test is the uniformly most powerful test of

$$H_0 : \mu_A = \mu_B \text{ versus } H_A : \mu_A \neq \mu_B.$$

Many of the standard statistical tests and estimators based on a population model are developed from the likelihood. We now show that the randomization mechanism is ancillary to the likelihood based on a population model. Let $t^{(j)} = (t_1, \ldots , t_j)$ and $y^{(j)} = (y_1, \ldots , y_j)$ be the realized treatment assignments and responses from patients $1, \ldots , j$, respectively. Let θ be the parameter of interest. Then the likelihood of the data after n patients, denoted as $\mathcal{L}_n$, is given by

$$\mathcal{L}_n = \mathcal{L}(y^{(n)}, t^{(n)}; \theta)$$
$$= \mathcal{L}(y_n|y^{(n-1)}, t^{(n)}; \theta)\mathcal{L}(t_n|y^{(n-1)}, t^{(n-1)}; \theta)\mathcal{L}_{n-1}. \quad (6.1)$$

Since the responses depend only on the treatment assigned and are independent and identically distributed under a population model, we have

$$\mathcal{L}(y_n|y^{(n-1)}, t^{(n)}; \theta) = \mathcal{L}(y_n|t_n; \theta). \quad (6.2)$$

Also, under complete or restricted randomization, the treatment assignments are independent of patient responses and consequently of θ (this will not be the case for response-adaptive randomization discussed in later chapters). Hence,

$$\mathcal{L}(t_n|y^{(n-1)}, t^{(n-1)}; \theta) = \mathcal{L}(t_n|t^{(n-1)}). \quad (6.3)$$

This likelihood reflects the specific restricted randomization procedure employed and the resulting dependence of t_n on $t^{(n-1)}$. Combining (6.1)–(6.3), we obtain

$$\mathcal{L}_n = \mathcal{L}(y_n|t_n; \theta)\mathcal{L}(t_n|t^{(n-1)})\mathcal{L}_{n-1}$$
$$= \prod_{i=1}^{n} \mathcal{L}(y_i|t_i; \theta)\mathcal{L}(t_i|t^{(i-1)}). \quad (6.4)$$

Since $\mathcal{L}(t_i | t^{(i-1)})$ is independent of θ, we have

$$\mathcal{L}_n \propto \prod_{i=1}^{n} \mathcal{L}(y_i | t_i; \theta). \qquad (6.5)$$

Note that the likelihood in (6.5) is identical to that arising from a nonrandomized design at fixed design points $t_1, \ldots, t_n$, that is, for any arbitrary sequence of treatment assignments, including nonrandom sequences. Consequently, a Bayesian[1], or believer in the likelihood principle could use (6.5) to justify an analysis ignoring the randomization mechanism. It is critical to point out that if we followed this approach to inference in this book, we could eliminate all chapters on inference, as the particular randomization procedure used would not matter in our analyses. Any biostatistics textbook would then cover the necessary population-based tests for clinical trials.

Unfortunately, clinical trials do not employ samples of patients that are drawn at random from infinitely large populations of patients on treatment A or B. In fact, there may be no patients at all on treatment A or B to sample from, if the treatments are experimental. Rather, patients are recruited into a clinical trial from various sources by a nonrandom selection of clinics in a nonrandom selection of locations. Clinics are selected because of their expertise, their ability to recruit patients, and their budgetary requirements. From these clinics, a nonrandom selection of eligible and consenting patients is performed, and these patients are then randomized to either treatment A or B.

Nevertheless, it has been argued that these samples of n_A and n_B patients individually are, in fact, representative of some larger undefined patient populations, even though they were not truly sampled at random. Arguing in this way, a population model can then be *invoked* as the basis for data analysis, with the assumption that $Y_{ij} \sim G(y | \theta_i)$. The invoked population model is shown on the right side in Figure 6.1. It is important to note that in performing the simplest t-test following a randomized clinical trial, a population model is being invoked.

Even if one could justify an invoked population model, we have discussed only *homogeneous* population models, where each patient is assumed to have the same underlying response distribution, depending only on the treatment assigned. Usually, the characteristics vary according to some underlying characteristics or vary over time. Therefore, even if patient selection for a trial could be viewed as representative sampling from an unspecified population, often the population would have to be viewed as heterogeneous. In the case of time heterogeneity, the underlying population model would likely have to incorporate changes over time in some unknown manner.

In conclusion, as stated by Lachin (1988, p. 296):

> The invocation of a population model for the analysis of a clinical trial becomes
> a matter of faith that is based upon assumptions that are inherently untestable.

---

[1]The Bayesian view on randomization is considerably more complex than this sentence suggests. The role of randomization in Bayesian inference will not be discussed in this book. Some excellent references are Rubin (1978) and Kadane and Seidenfeld (1990).

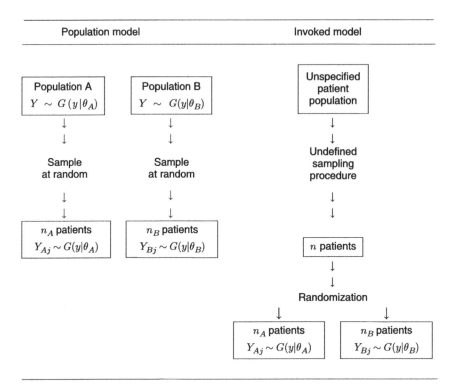

Fig. 6.1 *The population model versus the invoked population model for a clinical trial . Source: Lachin (1988, p. 295). Reproduced with permission of Elsevier.*

6.3 THE RANDOMIZATION MODEL

As we have seen in Section 6.2, due to the lack of a formal sampling basis, there is no formal statistical foundation for the application of population models to clinical trials. The randomization model is presented in Figure 6.2. Fortunately, the use of randomization provides the basis for an assumption-free statistical test of the equality of the treatments among the n patients actually enrolled and studied. These are known as *randomization tests*.

The null hypothesis of a randomization test is that the assignment of treatment A versus B had no effect on the responses of the n patients randomized in the study. This *randomization null hypothesis* is very different from a null hypothesis under a population model, which is typically based on the equality of parameters from known distributions. The essential feature of a randomization test is that, under the randomization null hypothesis, the set of observed responses is assumed to be a set of deterministic values that are unaffected by treatment. That is, under the null, each patient's observed response is what would have been observed regardless of whether treatment A or B had been assigned. Then the observed difference between the treatment groups depends only on the way in which the n patients were randomized. One

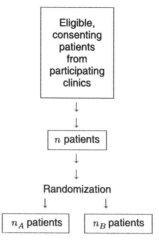

Fig. 6.2   *The randomization model for a clinical trial. Source: Lachin (1988, p. 295). Reproduced with permission of Elsevier.*

then selects an appropriate measure of the treatment group difference, or the treatment effect, which is used as the test statistic. The test statistic is then computed for all possible permutations of the randomization sequence. One then sums the probabilities of those randomization sequences whose test statistic values are at least as extreme as what was observed. This total is then the probability of obtaining a result at least as extreme as the one that was observed, which, by definition, is precisely the p-value of the test. A very small p-value (less than some α, say) then indicates that our observed value is quite extreme compared to the reference set of other possible randomization sequences and gives strong evidence to conclude that there is a difference between treatments. Randomization tests are assumption-free but depend explicitly on the particular randomization procedure used.

The simplicity of this approach to inference is often surprising to those rooted in the formal Neyman–Pearson theory of statistical hypothesis testing and demands some comment. In a sense, it is a direct contradiction to statistical hypothesis testing of a population parameter, because *here we treat the outcome variable of interest as fixed and the treatment assignments (design points) as random; in a population model, we traditionally treat the variable of interest as random at fixed values of the design points.* An easy way to think of this is as follows: patients enter the clinical trial with their outcome at the end of the trial prestamped on their foreheads, but the outcome is covered. After randomization, the cover is removed and the outcome noted. The only randomness is in the treatment assigned to the patient. If the null hypothesis is true, the outcome values should be evenly distributed across the As and the Bs.

Many statisticians have been convinced of the simple logic behind randomization tests. The following is a quotation attributed to Brillinger, Jones, and Tukey:

> If we are content to ask about the simplest null hypothesis, that our treatment has absolutely no effect in any instance, then the randomization, that must form

part of our design, provides the justification for a randomization analysis of our observed result. We need only choose a measure of extremeness of result, and learn enough about the distribution of the result for the observed results held fixed [and] for re-randomizations varying as is permitted by the specification of the designed process of randomization. If p percent of the values obtained by calculating as if a random re-randomization had been made are more extreme than (or equally extreme as) the value associated with the actual randomization, then p percent is an appropriate measure of the unlikeliness of the actual result. Under this very tight hypothesis, this calculation is obviously logically sound. [*Report of the Statistical Task Force to the Weather Modification Advisory Board*, 1978.]

However, a number of questions immediately arise. First, what measure of extremeness, or test statistic, should be used? The answer is any statistic that makes sense: a difference of means, a difference of proportions, a nonparametric rank test, even a covariate-adjusted treatment effect measure. A general family of statistics is based on the family of *linear rank tests* (e.g., Lehmann, 1986). Linear rank tests are used often in clinical trials, and the family includes such tests as the traditional Wilcoxon rank-sum test and the log-rank test, to name a few. We will focus extensively on linear rank tests in this book.

Second, which set of permutations of the randomization sequence should be used for comparison? If we use all possible permutations, the sequences *AAAA … AA* and *BBBB … BB* are included, and these sequences offer no information about the differences between treatments. In fact, if we used a randomization procedure that forces balances between treatments, shouldn't we compare to only those sequences with $n/2$ As and $n/2$ Bs? We will discuss the answer to this question in the next section.

Third, we have not discussed specific alternative hypotheses or error rates. Without error rates, how can one compute the power of the test? Power can only be determined under an invoked population model. However, power and sample size computations under a population model, as shown in Section 2.6, must be considered a crude approximation at best, with measures of variability determined in some sense by "best guesses." It is not unreasonable, therefore, to base sample size computations in planning a study on another "best guess"–an invoked population model, while still relying on a randomization test for analysis.

Fourth, if the analysis of a clinical trial is based on a randomization model that does not in any way involve the notion of a population, how can results of the trial be generalized to determine the best care for future patients? Berger (2000) argues that the difficulty in generalizing to a target population is a weakness not of the randomization test, but of the study design. If it were suspected by investigators that patient experience in a particular clinical trial could not be generalized, there would be no reason to conduct the clinical trial in the first place. Thus, we hope that the results of a randomized clinical trial will apply to the general population as well as to the patients in the trial. However, the study design only provides a formal assessment of the latter, not the former. By ensuring validity of the treatment comparison within the trial conducted, by limiting bias and ensuring strict adherence to the protocol, it is more likely that a generalization beyond the trial can be attained.

Lachin (1988) takes the approach that statistical inference in a clinical trial must be viewed as a two-step process. The first step is to determine whether there is a difference between treatments A and B among the n patients actually entered into the trial. The randomization test provides an assumption-free lock-tight test of this question. The second step is to ascertain the extent to which the observed results can be applied to an invoked population: the hypothetical population from which these n patients arose. For this, it is necessary to invoke a population model. However, this cannot be done with any statistical formalism. Rather, the only recourse is to precisely define the eligibility criteria adopted and then to present distributions of important baseline characteristics in order to describe the hypothetical population from which the study participants arose. The invoked population model then allows the construction of point estimates and confidence intervals and tests of the assumed population parameters.

Fifth, randomization tests are applicable to clinical trials where the primary interest is in demonstrating the superiority of one treatment over another via hypothesis testing. In *bioequivalence trials*, the goal is to demonstrate the equivalence of two treatments or noninferiority of one treatment over another. It is not clear how to conduct randomization tests to show bioequivalence, and this is a current research topic. There is also no analog between hypothesis testing and confidence intervals under a randomization model, which is very different from population-based inference where the acceptance region of a test can be analogous to a confidence interval around a parameter.

6.4   RANDOMIZATION TESTS

The *reference set* of a randomization test is the set of all permutations of randomization sequences that are used to evaluate the tail probability p-value in the comparison with our observed test statistic. An *unconditional reference set* is the set of all possible permutations, including those where all n assignments are to only one treatment A or B, or $n - 1$ to only one treatment, and so on. This is to be contrasted with a *conditional reference set*, which includes only those sequences with the same number of treatments assigned to A and B as were obtained in the particular randomization sequence employed. Let n_A be the observed number of patients assigned to treatment A, that is, the realization of $N_A(n)$. Let Ω be the cardinality of the reference set, with Ω_u and Ω_c the cardinality of the unconditional and conditional reference sets, respectively. The unconditional reference set will be substantially larger, having

$$\Omega_u = \sum_{n_A=0}^{n} \binom{n}{n_A} = 2^n$$

elements, while the conditional reference set will contain only

$$\Omega_c = \binom{n}{n_A}$$

elements. The conditional reference set is often favored in conjunction with procedures for which all 2^n sequences have positive probability (e.g., complete randomization, Efron's biased coin design (BCD), Smith's class of designs) because it excludes highly improbable sequences with large imbalances. As we have mentioned, the unconditional reference set contains two sequences (all As and all Bs) that give no information at all about treatment differences and many other sequences with large imbalances that have very little information about treatment differences. Using the conditional reference set is analogous to the traditional argument for conditioning in a population model wherein $N_A(n)$ is an ancillary statistic, which provides no information regarding the true treatment difference in the population. In later chapters, when we discuss response-adaptive randomization, $N_A(n)$ is no longer an ancillary statistic, and the same arguments do not apply.

When the random allocation rule (RAR) or the truncated binomial design (TBD) is used, there is no distinction between the unconditional and conditional reference sets, as we force $N_A(n) = n_A = n/2$. While the reference sets for the two designs are the same, the sequences in the reference sets have different probabilities. Also, the conditional reference set for complete randomization is equivalent to the reference set for the random allocation rule only on those occasions when we obtain $n_A = n/2$ following complete randomization.

Let S be the test statistic of interest, which can be any measure of the difference between the treatment groups. Define S_l to be the value of S for sequence $l, l = 1, \ldots, \Omega$ and define $S_{\text{obs.}}$ to be our observed test statistic. Let L record realizations of particular randomization sequences; L has a probability distribution depending on the particular randomization procedure employed. Then the p-value of the *unconditional randomization test* is given by

$$p_u = \sum_{l=1}^{\Omega_u} I(|S_l| \geq |S_{\text{obs.}}|) \Pr(L = l), \tag{6.6}$$

and the *conditional randomization test* is given by

$$p_c = \sum_{l=1}^{\Omega_c} I(|S_l| \geq |S_{\text{obs.}}|) \Pr(L = l|N_A(n) = n_A), \tag{6.7}$$

where $I(\cdot)$ is the indicator function. These p-values are two-sided.

One rejects the null hypothesis of no difference among the treatments for the n patients studied when $p < \alpha$ for some $\alpha \in (0, 1)$. The logic of including the observed sequence in the reference set has been argued extensively for exact tests, and compromises such as the mid p-value, where only half the probability of the observed sequence is included in the sum, have been suggested. The discreteness of the p-value becomes irrelevant as the sample size becomes larger. The reader is referred to Agresti (1990) for more details. On a related note, in small samples, the size of the reference set becomes important. Since, in small samples, the p-value cannot attain the exact alpha level, the larger the reference set, the more possible values the test will take on and more discrete values will be in the rejection region. Therefore, the test will become less conservative. Again, as the sample size becomes larger, this issue fades in importance.

We distinguish now between randomization tests and *permutation tests*, which are sometimes used interchangeably (including in the first edition of this book). However, it should be clear that when we are describing a randomization test, it is with respect to the reference set of all possible permutations of treatment assignments *produced by the particular randomization procedure employed*. A permutation test can be computed for any data set, randomized or not, provided the data are exchangeable. In the context of a permutation test, all permutations of the data are equiprobable. A *p*-value is computed in the same way, but every element of Ω_u has the same probability. In the context of a two-armed clinical trial, under complete randomization, a randomization test and permutation test are equivalent. Since permutational inference is valid under the exchangeability condition, why not just compute a permutation test, instead of incorporating the randomization procedure employed? In large samples, such tests may produce similar results. However, a basic principle of most statisticians is that "one should analyze as one designs." Restricted randomization can greatly reduce the size of the reference set. For example, a permuted block design (PBD) is very restrictive, and ignoring the blocks by simply conducting a randomization test as though complete randomization was used is inappropriate, just as ignoring the blocking factor in a population-based ANOVA model would be inappropriate.

6.5 LINEAR RANK TESTS

A special beauty of the randomization test is that any desired measure of a difference between treatments can be selected, such as difference of means or proportions, and so on. One never needs to formally describe the distribution or moments of a test statistic, as the randomization-based inference is completely determined by a *p*-value. However, this presumes that the entire reference set can be enumerated so that the randomization *p*-value can be computed. This is a tall order, even with today's computing, for even moderate sample sizes (see Section 6.8.1).

The family of linear rank tests is often used as the basis for a randomization test. Most standard nonparametric tests fall in this family, and these tests tend to have well-defined asymptotic distributions and, hence, can provide a nonparametric method for large-sample inference. Let $Y_1, \ldots, Y_n$ be the outcomes of the n patients and define a score function a_{jn} to be some score of the jth patient out of n patients, with arithmetic mean $\bar{a}_n$. Let $T_j, j = 1, \ldots, n$ be 1 if patient j was assigned to A and 0 if B. Then the linear rank statistic is given by

$$S = \sum_{j=1}^{n} (a_{jn} - \bar{a}_n) T_j. \tag{6.8}$$

The particular linear rank test is determined by the choice of a score function. For example, if the $\{a_{jn}\}$ are simple ranks,, the linear rank statistic is the well-known Wilcoxon rank-sum test.

Table 6.1   *Unconditional and conditional reference sets for computation of the linear rank test from complete randomization. Source: Lachin (1988, p. 298). Reproduced with permission of Elsevier.*

Unconditional ($\Omega_u = 16$)			Conditional ($\Omega_c = 6$)		
Sequence (l)	$\Pr(L = l)$	S_l	Sequence (l)	$\Pr(L = l)$	S_l
AAAA	1/16	0.0	AABB	1/6	−2.0
AAAB	1/16	−1.5	ABAB	1/6	0.0
AABA	1/16	−0.5	ABBA	1/6	1.0
AABB	1/16	−2.0	BAAB	1/6	−1.0
ABAA	1/16	1.5	BABA	1/6	0.0
ABAB	1/16	0.0	BBAA	1/6	2.0
ABBA	1/16	1.0			
ABBB	1/16	−0.5			
BAAA	1/16	0.5			
BAAB	1/16	−1.0			
BABA	1/16	0.0			
BABB	1/16	−1.5			
BBAA	1/16	2.0			
BBAB	1/16	0.5			
BBBA	1/16	1.5			
BBBB	1/16	0.0			

As an illustration, consider a clinical trial of four treatments with observed sequence *ABBA*. Suppose the patient outcomes were $Y_1 = 3$, $Y_2 = 1$, $Y_3 = 4$, $Y_4 = 5$. Then the simple ranks are $\{2, 1, 3, 4\}$. Under complete randomization, Table 6.1 gives a complete enumeration of the unconditional and conditional reference sets and the associated values of the Wilcoxon rank-sum test. From (6.6), we see that $p_u = 1/2$ and from (6.7), we see that $p_c = 2/3$.

Now suppose we were using Wei's UD(0, 1) design. Now the sequences are not equiprobable, and the respective probabilities are shown in Table 6.2. Here we compute $p_u = 1/2$ and $p_c = 1/2$.

For binary response data, one can assign binary scores $a_{jn} = 1$ or 0. Under a population model, the resulting linear rank test is algebraically equivalent to the usual Mantel–Haenszel chi-square test for the 2×2 contingency table under a conditional complete randomization model.

For survival data, the log-rank test can be obtained using *Savage scores* (Kalbfleisch and Prentice, 1980). In the usual notation of survival analysis, $\tau_1, \ldots, \tau_n$ are the event times of patients $1, \ldots, n$, and in the simplest case of no ties or censoring, we have n distinct ordered survival times $\tau_{(1)}, \ldots, \tau_{(n)}$ corresponding to treatment assignments $T_{(1)}, \ldots, T_{(n)}$. Then the linear rank statistic can be written as

$$S_n = \sum_{j=1}^{n} a_{(j)n} T_{(j)},$$

Table 6.2   *Unconditional and conditional reference sets for computation of the linear rank test from the* UD(0, 1). *Source: Wei and Lachin (1988, p. 352). Reproduced with permission of Elsevier.*

Unconditional ($\Omega_u = 16$)			Conditional ($\Omega_c = 6$)		
Sequence (l)	$\Pr(L = l)$	S_l	Sequence (l)	$\Pr(L = l)$	S_l
AAAA	0	0.0	AABB	0	−2.0
AAAB	0	−1.5	ABAB	1/4	0.0
AABA	0	−0.5	ABBA	1/4	1.0
AABB	0	−2.0	BAAB	1/4	−1.0
ABAA	1/12	1.5	BABA	1/4	0.0
ABAB	1/6	0.0	BBAA	0	2.0
ABBA	1/6	1.0			
ABBB	1/12	−0.5			
BAAA	1/12	0.5			
BAAB	1/6	−1.0			
BABA	1/6	0.0			
BABB	1/12	−1.5			
BBAA	0	2.0			
BBAB	0	0.5			
BBBA	0	1.5			
BBBB	0	0.0			

where

$$a_{(j)n} = 1 - \sum_{k=n-j+1}^{n} \frac{1}{k}, \tag{6.9}$$

which is equivalent to the log-rank statistic. We can also write the scores in (6.9) as

$$a_{jn} = 1 - E(X_{(j)}), \tag{6.10}$$

where $X_{(1)}, \dots, X_{(n)}$ are the order statistics from a unit exponential (Prentice, 1978; Kalbfleisch and Prentice, 1980, p. 79).

With censored data, let $C_1, \dots, C_n$ be the censoring times and $D_1, \dots, D_n$ be the event times of patients $1, \dots, n$. For the jth patient, we can only observe data pairs (Y_j, δ_j), where $Y_j = \min(D_j, C_j)$ and $\delta_j = I(D_j \leq C_j)$, where I is the indicator function. Assume that the censoring mechanism is the same in both treatment groups and that there are no ties. Let $\tau_{(1)} < \tau_{(2)} < \dots < \tau_{(M)}$ denote the M ordered distinct event times with R_m the number of study patients at risk just prior to $\tau_{(m)}$, $m = 1, \dots, M$. Then for the patient with an event at $\tau_{(m)}$, $\delta_j = 1$ and $Y_j = \tau_{(m)}$. For a censored patient, $\delta_j = 0$ and $\tau_{(m)} \leq Y_j < \tau_{(m+1)}$. Then for the log-rank test, the appropriate scores are given by

$$a_{(j)n} = 1 - \sum_{m=1}^{j} \frac{1}{R_m}$$

if $\delta_j = 1$ and

$$a_{(j)n} = -\sum_{m=1}^{j} \frac{1}{R_m}$$

if $\delta_j = 0$. If there are tied event times, the scores are calculated as though there were no ties, and then each of the patients with tied times is assigned the average of their scores.

6.6 VARIANCE OF THE LINEAR RANK TEST

The variance of the linear rank test can be described with respect to either the unconditional reference set or the conditional reference set of permutations. The unconditional variance of the linear rank test can be computed directly from (6.8) as

$$\mathrm{Var}(S) = \sum_{j=1}^{n} (a_{jn} - \bar{a}_n)^2 \mathrm{Var}(T_j)$$

$$+ \sum_{j=1}^{n} \sum_{\substack{i=1 \\ i \neq j}}^{n} (a_{in} - \bar{a}_n)(a_{jn} - \bar{a}_n)\mathrm{cov}(T_i, T_j). \qquad (6.11)$$

For complete randomization and the RAR, $\mathrm{Var}(T_j)$ and $\mathrm{cov}(T_i, T_j)$ do not depend on i or j, and (6.11) reduces to

$$\mathrm{Var}(S) = \{\mathrm{Var}(T_j) - \mathrm{cov}(T_i, T_j)\} \sum_{j=1}^{n} (a_{jn} - \bar{a}_n)^2. \qquad (6.12)$$

We can compute this quantity directly from Σ_T. For complete randomization, from (6.12), we have

$$\frac{1}{4} \sum_{j=1}^{n} (a_{jn} - \bar{a}_n)^2.$$

For the RAR, using (3.5) and (6.12), we obtain

$$\mathrm{Var}(S) = \frac{n}{4(n-1)} \sum_{j=1}^{n} (a_{jn} - \bar{a}_n)^2. \qquad (6.13)$$

The TBD variance is given by

$$\mathrm{Var}(S) = \frac{1}{4} \sum_{j=1}^{n} (a_{jn} - \bar{a}_n)^2$$

$$+ \frac{1}{2} \sum_{i=(n+2)/2}^{n} \Pr(\tau > i)(a_{in} - \bar{a}_n) \sum_{j=i+1}^{n} (a_{jn} - \bar{a}_n); \qquad (6.14)$$

the derivation is left as an exercise. For most other randomization procedures, an exact form of $\mathrm{Var}(S)$ is intractable, such as for Efron's BCD and Wei's urn design because the exact form of Σ_T is unknown.

The conditional variance with respect to the conditional reference set, is defined as $\mathrm{Var}(S|N_A(n) = n_A)$. Note that $\mathrm{cov}(T_j, T_{j'})$ is no longer 0 for complete randomization. In fact, $T_j, j = 1, \ldots, n$, are then dependent Bernoulli indicators with parameter n_A/n, so that $\mathrm{Var}(T_j) = n_A n_B/n^2$, where $n_B = n - n_A$. To find $\mathrm{cov}(T_i, T_j)$, we compute, for $j > i$,

$$
\begin{aligned}
E(T_i T_j) &= \Pr(T_i = 1, T_j = 1) \\
&= \Pr(T_j = 1 | T_i = 1)\Pr(T_i = 1) \\
&= \frac{n_A - 1}{n - 1}\left(\frac{n_A}{n}\right).
\end{aligned}
$$

Then

$$
\begin{aligned}
\mathrm{cov}(T_i, T_j) &= \frac{n_A - 1}{n - 1}\left(\frac{n_A}{n}\right) - \frac{n_A^2}{n^2} \\
&= -\frac{n_A n_B}{n^2(n - 1)}.
\end{aligned}
$$

Finally, we compute

$$
\mathrm{Var}(S|N_A(n) = n_A) = \frac{n_A n_B}{n}\left(\sum_{j=1}^{n} \frac{(a_{jn} - \bar{a}_n)^2}{(n - 1)}\right). \tag{6.15}
$$

Note that (6.15) reduces to (6.13) when $n_A = n/2$.

It is instructive to compare the variance of the linear rank test under a randomization model with the variance that would be obtained under a population model. For in a population model, one would assume that $t_1, \ldots, t_n$ are deterministic treatment indicators and $A_{1n}, \ldots, A_{nn}$ are independent and identically distributed random scores. The linear rank test under a population model can then be written as

$$
S_{\text{pop.}} = \sum_{j=1}^{n} t_j(A_{jn} - \bar{A}_n) = \sum_{j=1}^{n}\left(t_j - \frac{n_A}{n}\right)A_{jn}
$$

and

$$
\begin{aligned}
\mathrm{Var}(S_{\text{pop.}}) &= \mathrm{Var}(A_{jn})\sum_{j=1}^{n}\left(t_j - \frac{n_A}{n}\right)^2 \\
&= \frac{n_A n_B}{n}\mathrm{Var}(A_{jn}). \tag{6.16}
\end{aligned}
$$

We immediately see that the conditional linear rank test for complete randomization under a randomization model has a variance (6.15) that is a consistent estimator of the variance of the linear rank test under a population model, as given in (6.16). This observation gives more insight into the differences between the randomization and population models. If, in fact, we can assume that patient responses are independent and identically distributed, then, at least in very large trials using complete randomization or the RAR, the variance of the test statistic will be equivalent under the two models. However, other randomization procedures do not have this property, and as we have said, the assumption of a homogeneous population model may not be appropriate.

6.7 OPTIMAL RANK SCORES

In their classic text, Hájek and Šidák (1967) provide a general approach to the development of a nonparametric test for the comparison of two or more populations. They show the form of the optimal score generating function when it is desired to test the null hypothesis against a location or scale shift when sampling from a specific distribution. The resulting test is optimal in the sense of maximizing the Fisher's information in the data, and thus, it is asymptotically fully efficient. For example, simple rank (Wilcoxon) scores are optimal to detect a location shift when sampling from a logistic distribution, while van der Waerden scores (see Problem 7.4) are optimal for a normal distribution. Likewise, Savage scores are optimal to detect a scale shift in an exponential distribution. This theory was also used by Peto and Peto (1972) and by Prentice (1978) to derive the optimal rank scores for censored data under a proportional hazards and proportional odds alternative, the log-rank and modified Wilcoxon scores, respectively. Thus, there are a wide variety of score functions $\{a_{jn}\}$ that could be employed.

Under the randomization model, the responses are treated as fixed quantities, likewise the rank scores $\{a_{jn}\}$. Since the population model concept of sampling at random from two population distributions does not apply, the concept of efficiency does not strictly apply to the family of linear rank tests with a randomization-based distribution. However, one can still think about the average behavior of the test in repeated similar experiments. In this case, the expected properties of the observed responses might be relevant in the choice of the rank scores employed in the analysis. For example, if one thinks that the data from similar experiments are more likely to satisfy a proportional hazards alternative with censored data than a proportional odds alternative, one would choose to employ log-rank scores in the analysis rather than modified Wilcoxon scores.

For simple quantitative responses, there is a greater range of choices. Among these, simple rank scores are most commonly employed, in part due to the Mann–Whitney representation of the test, under a population model, as a function of $P(Y_A > Y_B)$ where Y_A represents a random observation from group A and Y_B likewise from group B. The Wilcoxon test statistic provides an estimate of this "proversion" probability, a useful quantity under a population model regardless of the underlying distributions.

While the "average" behavior of any one score function over a range of alternatives has not been thoroughly explored, it is reasonable to expect that the simple rank scores will yield a test that is in general robust to a location shift in any distribution.

Of course, the score function to be employed in any analysis must be prespecified. To compute multiple tests using different scores or to examine the properties of the data to choose the "best" score function would be cheating. Another approach would be to use a score-robust test that provides good power, in a population model sense, over a range of possible alternatives. One such test is the Gastwirth (1966) maximin efficient-robust test. This test is a convex combination of the standardized test (Z) values from the "extreme" pair in the set of tests considered. The extreme pair is determined by the estimated asymptotic relative efficiency of each pair of tests, which is equivalent to the square of the correlation of each pair. This approach could also be employed with the families of tests herein.

Consider two different tests using scores a_{jn} and b_{jn}, such as Wilcoxon scores and Savage scores. Let $a = (a_{1n}, \ldots, a_{nn})'$ and $b = (b_{1n}, \ldots, b_{nn})'$ refer to the corresponding vectors of scores and let Σ_T refer to the covariance matrix of the vector of treatment assignments $(T_1, \ldots, T_n)$. Then the covariance of the two test statistics is simply $a'\Sigma_T b$, from which the correlation between the two tests is obtained. Based on the resulting correlations between each pair of tests in the family of tests under consideration, the combination of tests is selected so as to maximize the minimum asymptotic relative efficiency relative to whichever test in the family would actually be optimal, if such were known in advance. See Lachin (2010) for the required expressions. This approach, however, is only applicable to those randomization designs for which the Σ_T is known explicitly. This includes complete randomization, the TBD, the RAR, and the PBD, and Efron's BCD.

6.8 EXACT AND LARGE-SAMPLE RANDOMIZATION TESTS

Ironically, the great founders of the randomized clinical trial understood the importance of randomization as a basis for inference, but were limited in their ability to perform it, due to the computational limitations of the day. But now that such tests can be performed using standard software packages in seconds, they are rarely used. While in the past, much of the literature was focused on finding the asymptotic distribution of the randomization test, such approximations were often inaccurate for moderate sample sizes, and the accuracy was highly dependent on the type of the randomization procedure employed. The computation of randomization tests using *Monte Carlo re-randomization* is now the preferred technique. For a fixed sequence of responses, one re-randomizes by generating L additional sequences, computing L tests based on the re-randomized sequences, and computes an estimated p-value. While this method is simple and relatively foolproof, there are some subtle considerations: what should L be, and how does one apply this technique with a conditional reference set? In this section, we compare methods for the computation of randomization tests for a test of the simple treatment effect.

6.8.1 Computation of exact tests

Even for conditional tests, enumerating all possible permutations in the reference set becomes prohibitively large as n gets larger than around 15. Mehta, Patel, and Wei (1988) provide a computational algorithm, which is effective for computing the exact distribution of randomization tests following restricted randomization procedures. Even so, such algorithms are probably only reasonable for sample sizes less than 50 unless parallel processing is used.

The basic networking algorithm is as follows. Let $P_{j+1}(n_A) = E(T_{j+1}|N_A(j) = n_{Aj})$, so that the algorithm applies to all restricted randomization procedures for which the $(j+1)$th treatment assignment depends on the previous treatment assignments only through $N_A(j)$ (this applies to the restricted randomization designs in Chapter 3). Let $T = (T_1, \dots, T_n)$ and let $\Omega_{n_A} = \{T : N_A(n) = n_{An}\}$. One does not have to enumerate every sequence in Ω_{n_A} in order to compute the exact distribution of the test statistic. The networking algorithm begins with a single node $(0, 0)$. For $j = 1, \dots, n-1$, each node (j, n_{Aj}) generates nodes $(j+1, n_{A,j+1})$ ending in a single terminal node (n, n_{An}). To each distinct subpath

$$(0,0) \rightarrow (1, n_{A1}) \rightarrow \dots \rightarrow (j, n_{Aj}),$$

assign a rank length

$$a_{1n}n_{A1} + a_{2n}(n_{A2} - n_{A1}) + \dots + a_{jn}(n_{Aj} - n_{A,j-1})$$

with associated probability

$$\prod_{k=1}^{j} \{P_k(n_{A,k-1})\}^{n_{Ak}-n_{A,k-1}} \{1 - P_k(n_{A,k-1})\}^{1-(n_{Ak}-n_{A,k-1})}.$$

Some of the rank lengths will not be unique; suppose there are $l(n_{Aj})$ distinct rank lengths, and denote them as $S_{jl}, l = 1, \dots, l(n_{Aj})$. Let π_{jl} be the sum of the probabilities for those paths that have the same rank length. Then the set $\Omega(j, n_{Aj}) = \{(S_{jl}, \pi_{jl}, l = 1, \dots, l(n_{Aj})\}$ is the probability distribution of $S_j = \sum_{i=1}^{j} a_{in}T_i$ given $N_A(j) = n_{Aj}$. One can then obtain each set $\Omega(j+1, n_{A,j+1})$ recursively from $\Omega(j, n_{Aj})$. The degree of computational efficiency gained by eliminating redundant sequences with the same rank length will of course depend on the scores a_{jn}. For the simple rank scores, this number increases as $O(n^2)$. However, for the log-rank test, the algorithm increases exponentially with n. Hollander and Peña (1988) provide an algorithm for exact enumeration when there are $K > 2$ treatments using Markov chain techniques.

In some special cases, a formula for the exact distribution of a randomization test can be derived. Galbete, Moler, and Plo (2015) provide the formula for binary responses under any restricted randomization procedure where ϕ_j is a function of only D_{j-1}.

6.8.2   Large sample randomization tests

Much of the literature in the 1980s on randomization-based inference involved computing the large sample distribution of the linear rank test under a randomization model. The linear rank test is particularly conducive to this type of asymptotic analysis. In general, one would presume that test statistics of the form

$$W = \frac{S}{\sqrt{\text{Var}(S)}}$$

should follow a standard normal distribution for large samples, based on our knowledge of the central limit theorem. In fact, this is only true when ϕ_j is a continuous function, so that procedures like Efron's BCD and the accelerated BCD do not admit asymptotically normal tests. The theory is complicated because the treatment assignments are correlated under the particular restricted randomization procedure. The interested reader is referred to the first edition of this book, where there are several chapters devoted to proving the asymptotic distribution of randomization tests.

The main condition for asymptotic normality is a Lindeberg-type condition on the scores (Smythe and Wei, 1983), requiring that

$$\lim_{n\to\infty} \frac{\max_{1\le j\le n}(a_{jn}-\bar{a}_n)^2}{\sum_{j=1}^n (a_{jn}-\bar{a}_n)^2} \to 0. \tag{6.17}$$

While this condition looks complicated, it essentially says that no individual absolute score can grow too large relative to the sum of all the absolute scores. For example, we could not use the actual data values from a continuous unbounded random variable. It is easy to see, for instance, that the simple ranks satisfy (6.17), as

$$\frac{\max_{1\le j\le n}(a_{jn}-\bar{a}_n)^2}{\sum_{j=1}^n (a_{jn}-\bar{a}_n)^2} = \frac{(n-1)^2}{\sum_{j=1}^n (2j-n-1)^2} = \frac{3(n-1)}{n(n+1)} \to 0$$

as $n \to \infty$, at a rate $O(1/n)$. Many other common score functions satisfy (6.17) as well (see Problem 6.8). If the scores satisfy the condition, then the unconditional test

$$W_U = \frac{2\sum_{j=1}^n (a_{jn}-\bar{a}_n)T_j}{\sqrt{\sum_{j=1}^n (a_{jn}-\bar{a}_n)^2}} \tag{6.18}$$

will be asymptotically standard normal. In particular, (6.18) is the correct form of the test statistic under complete randomization, the RAR, and Wei's urn design.

For the conditional test, conditional on $N_A(n) = n_A$, under complete randomization, the test statistic

$$W_C = \frac{\sum_{j=1}^n (a_{jn}-\bar{a}_n)T_j}{\sqrt{\gamma n_A(n-n_A)/n}}, \tag{6.19}$$

has an asymptotic standard normal distribution, where γ is defined by the additional assumption that

$$\gamma = \lim_{n\to\infty} \frac{\sum_{j=1}^{n}(a_{jn} - \bar{a}_n)^2}{n}. \tag{6.20}$$

For example, with continuous observations, defining the scores as $a_{jn} = r_{jn}/(n+1)$ where r_{jn} are the simple ranks, (6.20) is satisfied, and we have $\gamma = 1/12$ if there are no ties. Note that the addition to the sum of squares when there are ties is asymptotically negligible, provided the number of ties does not grow with n. However, this could be a problem for some outcomes, such as ordinal measures or continuous measures that are truncated to integer values (e.g., age). If there are many ties, one could substitute the observed value of (6.20) for γ in (6.19).

While these tests are simple to compute, under $UD(\alpha, \beta)$ randomization, the form of the conditional test statistic is more complicated (Smythe, 1988). Define a sequence of modified scores

$$b_{nn} = a_{nn} - \bar{a}_n,$$

$$b_{jn} = (a_{jn} - \bar{a}_n)$$

$$- \beta(2\alpha + (j-1)\beta) \sum_{k=j+1}^{n} \frac{(a_{kn} - \bar{a}_n)}{(2\alpha + (k-1)\beta)(2\alpha + (k-2)\beta)},$$

$$j = 1, \dots, n-1, \tag{6.21}$$

and let the $\{a_{jn}\}$ sequence be normalized so that $\sum_{j=1}^{n} b_{jn}^2 = 1$. Also define another sequence of modified scores, denoted as $\{\tilde{b}_{jn}\}$, which are computed by substituting $a_{jn} - \bar{a}_n = n^{-1/2}$ for all j into (6.21). Define

$$\rho_n = \sum_{j=1}^{n} b_{jn}\tilde{b}_{jn}, \quad s^2 = \lim_{n\to\infty} \sum_{j=1}^{n} \tilde{b}_{jn}^2.$$

Then, conditional on $D_n = N_A(n) - N_B(n) = d_n$, the test is given by

$$W_C = \frac{2\sum_{j=1}^{n}(a_{jn} - \bar{a}_n)T_j - \rho_n d_n/(\sqrt{n}s^2)}{\sqrt{1 - \rho_n^2/s^2}}.$$

While it may be tempting to employ the form of the test in (6.19), in this case, simulations show that the test is slightly anticonservative, and the more complicated form is more appropriate.

6.9   MONTE CARLO RE-RANDOMIZATION TESTS

In this section, we describe how to compute unconditional and conditional tests using Monte Carlo re-randomization. The techniques can be implemented in SAS or R and run very quickly.

6.9.1   Unconditional tests

The general algorithm for an unconditional Monte Carlo re-randomization test is as follows. For a set of observed responses $x_1, \ldots, x_n$ and the treatment assignments used in the trial $t_1, \ldots, t_n$, generated by a randomization procedure ϕ_j, we compute a test statistic, which can be based on any treatment effect difference, and call it $S_{\text{obs.}}$. Now we generate L randomization sequences using Monte Carlo simulation. For each of these sequences, a new test statistic, $S_l, l = 1, \ldots, L$, is computed from $x_1, \ldots, x_n$. The two-sided Monte Carlo p-value estimator is then defined as

$$\hat{p}_u = \frac{\sum_{l=1}^{L} I(|S_l| \geq |S_{\text{obs.}}|)}{L}. \tag{6.22}$$

For restricted randomization, the key component of this computation is that disparate probabilities of sequences will be depicted by the frequency of duplicate sequences sampled with replacement.

Whether or not S_l is extreme, is distributed as Bernoulli with underlying probability p_u, and hence, $\hat{p}_u$ is unbiased with

$$\text{MSE}(\hat{p}_u) = \frac{p_u(1 - p_u)}{L}.$$

Then establishing a bound $\text{MSE}(\hat{p}_u) < \epsilon$ implies that $L > 1/4\epsilon$. For $\epsilon = 0.0001$, we have $L > 2500$ (Zhang and Rosenberger, 2011). The value of ϵ may not be small enough to estimate very small p-values accurately. Plamadeala and Rosenberger (2012) suggest finding L that ensures $P(|\hat{p}_u - p_u| \leq 0.1p_u) = 0.99$, for instance. It follows that $L \approx (2.576/0.1)^2(1 - p_u)/p_u$. Thus, to estimate a p–value as large as 0.04 with an error of 10% with 0.99 probability, the Monte Carlo sample size must be $L = 15,924$. If a smaller p-value is expected, L will be larger. In any event, generating 20,000 randomization sequences takes only seconds. Galbete and Rosenberger (2015) demonstrate by simulation that 15,000 sequences produce tests that are almost identical in distribution to exact tests (when exact tests can be computed directly) and are considerably more accurate than asymptotic tests for moderate sample sizes.

The method of Monte Carlo re-randomization is so simple that it often invites skepticism. There are indeed subtle considerations, as in any inference procedures. While in the population model we are concerned about assumptions of the model and which test statistic is appropriate, these are not concerns in the randomization model approach. Instead, we are concerned about the appropriate reference set. In Section 6.9.3, we describe tests computed under the conditional reference set, which many people prefer, for reasons described in Section 6.4, but are considerably more complicated than conditional tests. In Chapter 7, we describe stratified versus unstratified tests.

In the analysis of variance context, the use of the PBD requires the incorporation of a blocking factor to provide valid inference (cf. Matts and Lachin, 1988). The analog under a Monte Carlo re-randomization test is not obvious. One possible approach would be to compute a separate test within each block and combine tests by summing over blocks, analogous to a stratified test (see Section 7.5). This would adjust for any intrablock correlation that may be present. However, block sizes may be small, and there may be an incomplete block; any incomplete block with only one patient would have to be discarded. Such a blocked analysis would require the observed block sizes to be fixed in a re-randomization procedure. However, the randomizing of the blocks is as important a component of the randomization procedure as the treatment assignments themselves. A re-randomization analysis therefore produces a reference set that is representative of the randomization procedure actually employed, both re-randomizing the block sizes and the patients within blocks, and is consistent with the methodology presented for all restricted randomization procedures.

6.9.2 Example

In Appendix A, we have data on cholesterol levels for 50 patients from the Diabetes Complications and Control Trial (DCCT), and the randomization sequence, which was generated using Wei's UD(0, 1) procedure. In Table 6.3, we compute the Monte Carlo re-randomization test unconditional p-value estimate, assuming that complete randomization, Efron's BCD, or Wei's UD(0, 1), or the big stick design ($b = 3$) was used. Only the p-value for Wei's UD(0, 1) is correct, because that was the procedure actually used in the trial, although the others are close. The big stick design is different because the observed sequence could not have been generated under that design, since there are imbalances greater than 3. This demonstrates the importance of using the correct reference set. We use $L = 20,000$ as the Monte Carlo sample size. The test statistic is the linear rank test with simple rank scores; the value of $S_{obs.}$ is 29.5.

It is not difficult to find examples of sequences in very small trials for which the p-values will not be the same if different randomization procedures are used (see Problem 6.11). The ease of re-randomizing from the correct randomization procedure

Table 6.3 *Monte Carlo re-randomization test p-values for the DCCT data in Appendix A under four different re-randomization procedures.*

Procedure	p-value
Complete randomization	0.5647
BCD ($p = 2/3$)	0.5665
UD(0, 1)	0.5665
Big Stick ($b = 3$)	0.5850

makes it rather silly not to use the test based on the actual randomization procedure employed, even if the p-values are nearly identical. The SAS program to generate the randomization test for the UD(0, 1) procedure is given in Appendix B, and it takes less than a second with $L = 20,000$.

6.9.3 Conditional tests

The simple approach to generating conditional re-randomization tests is outlined by Zhang and Rosenberger (2011): to generate a very large number of sequences, say $K \gg L$, and then keep only those sequences that satisfy the condition $N_A(n) = n_A$. Suppose that at least N_c sequences satisfying $N_A(n) = n_A$ are sufficient to approximate the conditional randomization distribution of the test. Let K sequences be sampled, $\mathbf{T}_1, \dots, \mathbf{T}_K$, independently and with replacement from the unconditional reference set using $\phi_j = E(T_j | \mathcal{F}_{j-1})$ as the sampling mechanism. The requisite number of sequences, K, follows a negative binomial random variable with parameters $\pi = P(N_A(n) = n_A)$ and $r = N_c$ (Zhang and Rosenberger, 2011). Let N denote a value in the range of K, $N = N_c, N_c + 1, \dots$. For $l = 1, \dots, N$, a sequence $\mathbf{T}_l$ is sampled from the unconditional reference set. The lth sampled sequence induces two Bernoulli random variables

$$Y_l = \begin{cases} 1, & \text{if } N_A(n) = n_A \\ 0, & \text{otherwise}, \end{cases}$$

and

$$X_l = \begin{cases} 1, & \text{if } N_A(n) = n_A \text{ and } |S_l| \geq |S_{\text{obs.}}|, \\ 0, & \text{otherwise}. \end{cases}$$

A consistent estimator for the p-value of the conditional test can be computed as

$$\hat{p}_c = \frac{\sum_{l=1}^{N} X_l}{\sum_{l=1}^{N} Y_l}.$$

Plamadeala and Rosenberger (2012) show that this naive approach is computationally infeasible and gets rapidly more complex as n_A deviates from 0.5. For instance, if Efron's BCD is used with $p = 0.75$, $n = 200$, and $n_A = 90$, the naive approach requires $K = 6,754,269 \times 10^6$ to produce $N_c = 2500$ sequences. They present an alternate approach, by sampling directly from the conditional reference set.

To guarantee a sequence from Ω_c, T_j in ϕ_j must be conditioned on both $N_A(j)$ and $N_A(n)$. Consequently, for $0 \leq m_j \leq j$, the procedure

$$p_j = \begin{cases} P(T_j = 1 | N_A(j-1) = m_{j-1}, N_A(n) = n_A), & 2 \leq j \leq n, \\ P(T_j = 1 | N_A(n) = n_A), & j = 1, \end{cases} \tag{6.23}$$

must be applied to generate a random sequence strictly from Ω_c. A general formula links ϕ_j to the new randomization procedure p_j:

$$p_j = \begin{cases} \phi_j \dfrac{P(N_A(n)=n_A|N_A(j)=m_j)}{P(N_A(n)=n_A|N_A(j-1)=m_{j-1})}, & 2 \leq j \leq n, \\[2ex] \phi_j \dfrac{P(N_A(n)=n_A|T_j=1)}{P(N_A(n)=n_A)}, & j = 1. \end{cases} \tag{6.24}$$

One therefore needs to compute the conditional probabilities of the form

$$P(N_A(n) = n_A|N_A(j) = m_j),$$

and plug them into (6.24) and re-randomize N_c sequences using p_j as the randomization procedure instead of ϕ_j.

In the simplest case of complete randomization,

$$p_j = \frac{n_A - m_{j-1}}{n - (j-1)}.$$

If $n_A = n/2$, this is the RAR. Plamadeala and Rosenberger (2012) compute p_j for Efron's BCD. All other restricted randomization procedures should be considered open problems.

6.10 PRESERVATION OF ERROR RATES

If a population model is assumed, a parametric test based on that model should be more powerful under that model than a nonparametric alternative. This is a well-known fact, and it should also apply to randomization tests. However, as we have stated, the invocation of a population model involves untestable assumptions. Therefore, even the use of the common t-test cannot be justified under a randomization model. In this section, we simulate the error rates under a population model. Galbete and Rosenberger (2015) simulated the behavior of the t-test and re-randomization test under a linear time trend. The linear time trend violates the assumptions of the t-test, making the variance of the test and the estimated treatment effect wrong. However, the randomization test should be robust to time trends. Table 6.4 gives the error rates for the randomization test and the t-test when responses are assumed to be normally distributed, for $n = 50$. The test statistic for the randomization test is the difference of means. We investigate two models:

1. Under H_0, $X_1, \ldots, X_n \sim$ i.i.d. $N(0, 1)$.
   Under H_1, treatment A has a mean shift of 1.
2. Under H_0, $X_1, \ldots, X_n$ are subject to a drift over time, ranging linearly on the interval $(-2, 2]$ plus a $N(0, 1)$ random variable.
   Under H_1, treatment A has a mean shift of 1.

Table 6.4 Simulated size and power of the randomization test and the t-test under two different models and seven different randomization procedures. Each simulation based on 10,000 tests, n = 50. For the randomization test, L = 15,000. All tests are unconditional unless noted.

| | Model (1) | | | | Model (2) | | | |
| | Randomization | | t-Test | | Randomization | | t-Test | |
Procedure	Size	Power	Size	Power	Size	Power	Size	Power
CR	0.05	0.87	0.05	0.93	0.05	0.57	0.05	0.60
CR-cond.	0.05	0.87			0.05	0.57		
RAR	0.04	0.93	0.04	0.93	0.05	0.61	0.04	0.60
TBD	0.05	0.93	0.05	0.93	0.05	0.35	0.18	0.57
UD(0, 1)	0.05	0.91	0.05	0.93	0.05	0.66	0.02	0.62
BCD	0.04	0.92	0.05	0.93	0.05	0.78	0.01	0.64
PBD	0.05	0.93	0.04	0.93	0.05	0.88	0.00	0.65
RBD	0.05	0.93	0.04	0.93	0.05	0.90	0.00	0.65
BSD	0.05	0.93	0.05	0.93	0.05	0.83	0.00	0.61

The nominal size should be 0.05 and the nominal power of the t-test under model (1) should be around 0.93. Under any model, a large linear time trend should necessarily reduce power. We investigate the impact of eight different randomization procedures: complete randomization (both unconditional and conditional tests), RAR, TBD, Wei's UD(0, 1), Efron's BCD (BCD, $p = 2/3$), the big stick design (BSD) with $b = 3$, the PBD with $m = 4$, and the random block design (RBD) with $B_{max} = 3$. Each test was simulated 10,000 times (for each of the 10,000 tests, both the response data and treatment assignments were regenerated), and each re-randomization test was computed using $L = 15,000$ sequences.

Both tests maintain size and power when there is no time trend (Model (1)) with only one exception: the t-test is more powerful than the randomization test under complete randomization, for both the unconditional and conditional tests. When we incorporate a very strong linear time trend, the t-test fails to maintain nominal size, except under complete randomization and the RAR. This phenomenon was also observed by Rosenkranz (2011). However, under every randomization procedure, the randomization test preserved the type I error rate exactly. This is the hallmark of the randomization test: under model misspecification, the randomization test protects the type I error rate. Time trends often are hidden from the investigators and cannot be perceived in advance of the trial (Altman and Royston, 1988). Yet investigators can be assured that the type I error rate of the trial will not be affected.

We included the TBD in the table because of our discussion from Chapter 4 that the accidental bias increases with n and that the probability of covariate imbalance is large in the presence of time trends. Here we see that time trends greatly reduce the power of both the randomization test and t-test, but that the type I error rate is greatly inflated for the t-test, while the randomization test has nominal size. This is the only randomization procedure for which the t-test has inflated size.

The t-test is better than the randomization test, in the presence of a time trend, only under complete randomization, with slightly higher power. For all restricted randomization procedures that are not forced balance designs, power is increased by 4–35 percent by using a randomization test. The RBD is particularly effective in protecting against time trend, with only 3 percent loss in power. We conclude that randomization tests should be used as a matter of course to protect the type I error and improve power when a hidden time trend is present unless complete randomization is used.

6.11   REGRESSION MODELING

Thus far, we have described randomization tests for the simple hypothesis test of the treatment effect. However, it may be desired to test a covariate-adjusted treatment effect from a regression model. While a regression model is usually developed under a population model, it is straightforward to apply a randomization analysis following the fit of a model. Conceptually the basic steps are to first fit a model to baseline covariates, other than treatment group. Then the residuals from the model can be viewed as a set of preordained responses, regardless of which treatment is assigned. The model residuals can then be employed in lieu of the responses as the basis for computing a rank score. This approach was first described by Gail, Tan, and Piantadosi (1988). They used the asymptotic distribution of the randomization test assuming that treatment assignments are equiprobable.

A major advantage of the randomization analysis of the residuals is that the validity of the test in no way depends on the validity of the model assumptions used to fit the model. Thus, if a simple normal errors model is used as the basis for computing the residuals, then the validity of a t or F-test between groups depends on the homoscedastic normal errors assumption. However, the randomization test comparing the randomly assigned groups in no way depends on this assumption. Thus, the randomization test can be viewed as a robust test in situations where the regression model may be misspecified.

Under the generalized linear model (GLM) formulation of McCullagh and Nelder (1989), let Y be the outcome variable, X the covariate, and T the treatment indicator, which is independent of X; and let $h(\cdot)$ be a known function, and η be the parameter representing the linear relationship. The GLM is given as follows:

$$\eta = \mu + T\alpha + X\beta;$$

$$E(Y|X, T) = h(\eta).$$

Define $\theta(\cdot)$ to be the canonical parameter of an exponential family with first derivative $\theta'(\cdot)$, ϕ as a scale parameter, $d(\cdot)$ and $b(\cdot)$ as functions identifying the distribution of Y, and $R(\cdot)$ is a function that does not contain any unknown parameters. The log-likelihood of Y conditional on X and T is as follows:

$$\ell = \{d(\phi)\}^{-1}[Y\theta(\eta) - b\{\theta(\eta)\} + R(Y)] + c(Y, \phi).$$

The regression model is thus $E(Y|X, T) = h(\eta) = h(\mu + T\alpha + X\beta)$, where $h(\eta) = b'(\theta(\eta))$ and $b'(\cdot)$ is the first derivative. Under the null hypothesis H_0 of the randomization test, there is no difference between treatment groups A and B. Gail, Tan, and Piantadosi (1988) proposed to construct randomization tests based on the randomization distribution of the residuals r from the fitted model obtained from maximum likelihood estimation. Suppose ϕ is known, let $\hat{\eta}_0 = \hat{\mu}_0 + X\hat{\beta}_0$, where $\hat{\mu}_0$ and $\hat{\beta}_0$ represent maximum likelihood estimators of μ and β under the null hypothesis $\alpha = 0$. The score function then is

$$U = \frac{\partial \ell}{\partial \alpha}$$

$$= \{d(\phi)\}^{-1} \sum \left(Y \frac{\partial \theta}{\partial \eta} \frac{\partial \eta}{\partial \alpha} - \frac{\partial b}{\partial \theta} \frac{\partial \theta}{\partial \eta} \frac{\partial \eta}{\partial \alpha} \bigg| \eta = \hat{\eta}_0 \right)$$

$$= \{d(\phi)\}^{-1} \sum T\theta'(\hat{\eta}_0)\{Y - h(\hat{\eta}_0)\}.$$

The residuals r have the following form:

$$r = (d(\phi))^{-1}\theta'(\hat{\eta}_0)(Y - b'(\theta(\hat{\eta}_0))) = (d(\phi))^{-1}\theta'(\hat{\eta}_0)(Y - h(\hat{\eta}_0)).$$

Parhat, Rosenberger, and Diao (2014) directly compute the randomization test by ranking the residuals and calculating the linear rank test, then re-randomizing for both unconditional and conditional tests, using the techniques in Section 6.9. They also apply the technique to time-to-event data using both the Cox proportional hazards model and the accelerated failure time model, by ranking the martingale residuals (cf. Fleming and Harrington, 1991). They find that the randomization test preserves error rates better than tests based on the population model, when the underlying model is misspecified.

For longitudinal outcomes, where the rate of change of an outcome over time is of interest, Parhat, Rosenberger, and Diao (2014) compute a randomization test based on the predictors of random slopes from a generalized linear mixed model.

6.12 ANALYSES WITH MISSING DATA

An important concern in both randomization-based and population-based inference is the validity of an analysis when responses $\{Y_j\}$ for some patients are missing. Often patients with missing data are simply excluded and the analysis is based only on the subset of patients with complete data. Under a population model, such an analysis can be justified when it can be assumed that the missing data are missing completely at random (MCAR), or that missingness is statistically independent of the observable response (cf. Little and Rubin, 1987).

Let Y_j be the potentially observable response, and let v_j be an indicator variable to denote whether the response of the jth patient is observed ($v_j = 1$) or missing ($v_j = 0$). Since we will only conduct an analysis within the subgroup with observed data, only one subgroup indicator variable is employed. Under a population model, the treatment assignments T_j are deterministic, and under the MCAR assumption, the responses $\{Y_j\}$ are statistically independent of the $\{v_j\}$. Under a randomization model, however, the responses Y_j are fixed and it is assumed that the treatment assignments $\{T_j\}$ are statistically independent of the $\{v_j\}$ (see Lachin, 1988). This implies that the expected probability of treatment assignment is the same for those patients with observed data and those with missing data. This condition is a special case of the *covariate independence assumption*, to be described in Section 7.5.

Unfortunately, under either a population model with the MCAR assumption, or a randomization model with the independence assumption, there is no possible direct test for the required assumptions. These assumptions can only be assessed indirectly, such as by examining the observed characteristics of those patients with missing data versus those with observed data. For this reason, it is important that the incidence of missing data be kept to a minimum.

Linear rank tests are particularly conducive to analyses involving missing data under a randomization model. There are basically two approaches: the *complete data analysis* and the *worst rank analysis*. Both procedures are compared in Kennes, Hilgers, and Heussen (2012), but they were unable to draw any global conclusions on the desirability of either method.

For the linear rank test, when no observations are missing, the rank score a_{jn} can be written as some function $a_{jn} = f(Y_1, \ldots , Y_n)$. In the presence of missing data, however, the rank score is defined as $c_{jn} = f(v_1, Y_1, \ldots , v_n, Y_n)$; that is, c_{jn} is defined as a function of the complete observations only and is undefined if Y_j is missing. The linear rank test then becomes $S = \sum_{j=1}^{n} v_j(c_{jn} - \bar{c}_n)T_j$ where $\bar{c}_n = (\sum_{j=1}^{n} v_j c_{jn})/(\sum_{j=1}^{n} v_j)$. An analysis using this test is called a complete data analysis

In some instances, response measures are informatively missing for reasons that are obviously not random in a population model sense or independently in a randomization model sense. A common example is where the outcome measure $\{Y_j\}$ is a measure of quality of life or disease severity but where some subjects die during the study. In this case, the fact that the subject died indicates the worst possible quality of life or disease severity. For such cases, Lachin (1999) describes a worst-rank analysis in which subjects who die are assigned a common value more extreme than that of the observed values, such that all deaths share a tied worst rank. Lachin (1999) also describes a method for assigning rank scores in a time to event analysis such that the deaths have untied worst scores.

6.13   SAMPLE SIZE CONSIDERATIONS FOR RANDOM SAMPLE FRACTIONS

In Section 2.6, we described the calculation of appropriate sample sizes under a population model. There is no analog in randomization-based inference. Since many components of the sample size considerations are "best guesses" before the trial is conducted, in practice, we may use the sample size computed under a "best guess" population model as an approximation of the sample size required for randomization-based inference. Formula (2.9) assumes that the sample fractions

Table 6.5 *Sample size requirements for Smith's design for* $\rho = 0, 1, 5$ *($\alpha = 0.05$, $\beta = 0.8$, and $\mu_A - \mu_B = 1$). Source: Hu and Rosenberger (2006, p. 101), reproduced with permission of John Wiley and Sons, Inc.*

(σ_A, σ_B)	ρ	Type I	Type II	Type III
(1, 1)	0	25	26	28
	1	25	26	25
	5	25	25	25
(1, 2)	0	62	63	72
	1	62	63	68
	5	62	65	65
(1, 4)	0	211	212	234
	1	211	212	223
	5	211	211	217

are fixed, but in complete randomization and restricted randomization designs that are not forced-balance, they are random. Hu and Rosenberger (2006) describe formulas such as (2.9) as yielding a *type I sample size* estimate. When $N_A(n)$ and $N_B(n)$ are random variables, they discuss two other types of sample size estimation: *type II sample size estimation*, which is the average sample size computed over the distribution of $N_A(n)$; and *type III sample size estimation*, which is computed over a large percentile (say 90th or 95th), of the distribution of $N_A(n)$. While most statisticians might naively assume fixed sample fractions even if they are random, even taking a mean over the distribution of $N_A(n)$ ensures that only 50 percent of clinical trials will attain the requisite power. Since we only run a clinical trial once, the idea of computing a sample size at a high percentile of the sample size distribution ensures that the probability of attaining the requisite power is high.

Hu and Rosenberger (2006) use the asymptotically normal distribution of restricted randomization procedures (where they exist) to compute the type I (formula (2.9)), type II, and type III sample sizes. Table 6.5 gives the required sample size for a test with $\alpha = 0.05$, $\beta = 0.8$, and a treatment effect of 1 under Smith's design ($\rho = 0, 1, 5$) when the variances in each group are equal and unequal. Note that the type III sample size for complete randomization ($\rho = 0$) is considerably larger than the type I sample size when variances are unequal, due to the large variability of $N_A(n)$, but it is only slightly higher for $\rho = 5$. For equal variances, all three sample size estimates are the same for Smith's design with $\rho = 5$. The conclusion is that type III sample size estimation should be used when there are random sample fractions and that it may lead to a slightly increased sample size requirement unless the variability of the procedure is small.

6.14 GROUP SEQUENTIAL MONITORING

In many clinical trials, it is desirable to establish a sequential monitoring plan, whereby the test statistic is computed at an interim point or points in the trial and a decision is made whether to stop early due to evidence of treatment efficacy, while preserving the overall type I error rate. When the test statistic is computed and decisions are made after groups of patients have responded to treatment, such a plan is called *group sequential monitoring*. There is a large literature on group sequential monitoring of population-based inference procedures; see Jennison and Turnbull (2000) for a comprehensive overview of the subject. In this section, we describe a group sequential monitoring strategy for monitoring randomization tests.

6.14.1   Establishing a stopping boundary

Here, consider just a single interim monitoring point after n_1 patients have responded; the basic formulation can be extended to any number of inspections. Let

$$S_{n_1} = \sum_{j=1}^{n_1} (a_{jn_1} - \bar{a}_{n_1})T_j = a'_{n_1}T_{n_1}$$

be the computed linear rank statistic after n_1 patients and let

$$S_n = \sum_{j=1}^{n} (a_{jn} - \bar{a}_n)T_j = a'_n T_n$$

be the computed statistic at the end of the trial, where $a'_{n_1} = (a_{1n_1} - \bar{a}_{n_1}, \ldots, a_{n_1 n_1} - \bar{a}_{n_1})$, $a'_n = (a_{1n} - \bar{a}_n, \ldots, a_{nn} - \bar{a}_n)$, $T'_{n_1} = (T_1, \ldots, T_{n_1})$, and $T'_n = (T_1, \ldots, T_n)$.

It is necessary to find the joint probability distribution of (S_{n_1}, S_n). This could be computed exactly, as in Lin, Wei, and DeMets (1991), or using the asymptotic joint distribution. The *spending function approach* of Lan and DeMets (1983) requires specifying a strictly increasing continuous function $\alpha^*(t), t \in [0, 1]$ such that $\alpha^*(0) = 0$ and $\alpha^*(1) = \alpha$. Then we find constants d_1 and d_2 such that

$$\Pr(S_{n_1} > d_1) = \alpha^*(t_1)$$

and

$$\Pr(S_n > d_2, S_{n_1} \le d_1) = \alpha^*(1) - \alpha^*(t_1), \tag{6.25}$$

where t_1 represents the fraction of information available, with respect to total information accrued in the entire clinical trial, after n_1 patients.

In the usual case where the test statistic is a sum of independent and identically distributed random variables, we could write $S_n = S_{n_1} + S_{n_2}$, where S_{n_1} and S_{n_2} are independent. This simplifies the distribution theory dramatically. Unfortunately, for the linear rank test, the first n_1 elements of $\{a_{jn}\}$ are not necessarily the same as $\{a_{jn_1}\}$, and hence, the linear rank statistic cannot be decomposed. Even if it could be decomposed in this way, S_{n_1} and S_{n_2} would not be independent, except under complete randomization.

If unconditional inference is used, we can easily compute the covariance of the test statistics as follows. Let $\Sigma_{n_1} = \mathrm{Var}(\boldsymbol{T}_{n_1})$ and $\Sigma_n = \mathrm{Var}(\boldsymbol{T}_n)$ Assuming that (S_{n_1}, S_n) are jointly asymptotically normal and Σ_n does not depend on n, we can see that $\mathrm{Var}(S_{n_1}) = \boldsymbol{a}'_{n_1}\Sigma_{n_1}\boldsymbol{a}_{n_1}$, $\mathrm{Var}(S_n) = \boldsymbol{a}'_n\Sigma_n\boldsymbol{a}_n$ and

$$\mathrm{cov}(S_{n_1}, S_n) = [\boldsymbol{a}'_{n_1} : \boldsymbol{0}_{1\times n-n_1}]\Sigma_n\boldsymbol{a}_n. \tag{6.26}$$

The probabilities in (6.25) can be computed using this covariance under the correct asymptotic joint distribution.

For conditional inference, the variance–covariance structure of $T_1, \ldots, T_{n_1}, \Sigma_{n_1}$, will be determined conditional on $N_A(n_1) = n_{A1}$, say, and this will no longer be a sub-matrix of Σ_n, since it will be a function of n_{A1}. In this case, the variance–covariance matrix of treatment assignments among the first n_1 allocations with the vector of n allocations is more complicated than for the unconditional test.

Finding the joint asymptotic distribution of sequentially computed linear rank statistics under different randomization procedures using given score functions is an open topic. From (6.26), it is clear that in many cases the test statistics do not have independent increments.

Plamadeala and Rosenberger (2012) develop a technique to compute the Monte Carlo conditional distribution of sequentially computed test statistics when Efron's BCD is used. This allows for the monitoring of K sequentially computed conditional randomization tests at interim inspections.

6.14.2 Information fraction

The spending function approach allows for any arbitrary sequence of interim inspections, and unplanned inspections can be added by simply computing $\alpha^*(t)$ for the given information at the additional interim inspection. For the spending function approach, we must determine the fraction of total information accrued in the trial after n_1 patients. Under a population model, the information attained by n_1 would be defined according to the Fisher's information for the estimator of the parameter of interest, which is asymptotically equivalent to the inverse of the variance of the estimator (to first order). For an estimation-based test, constructed as the ratio of the estimator to its standard error, then the variance of the test decreases in n and information increases. However, for such tests expressed as a partial sum rather than a mean, the expression for the information is proportional to the variance of the sum, with both increasing in n.

Since the linear rank test involves the sum of the scores, this suggests that the variance of the test could be used as a measure of the information in the data. Thus, even though we are not operating under a population model, the information fraction at n_1 is given by

$$t_1 = \frac{\boldsymbol{a}'_{n_1}\Sigma_{n_1}\boldsymbol{a}_{n_1}}{\boldsymbol{a}'_n\Sigma_n\boldsymbol{a}_n}.$$

In general, while Σ_n will be known for some randomization procedures, we will not know $\boldsymbol{a}_n$. Thus, it is necessary to employ a surrogate measure of information, as in

Lan and Lachin (1990). Let $n_A(t_1)$ and $n_B(t_1)$ be the number of patients assigned to treatment A and B, respectively, at the time of the interim inspection, and let $n(t_1) = n_A(t_1) + n_B(t_1)$. Since the trial will have a target sample size in each group, n_A and n_B, where $n = n_A + n_B$, then one possible surrogate could be

$$\tilde{t} = \frac{\frac{n_A(t_1)n_B(t_1)}{n(t_1)}}{\frac{n_A n_B}{n}} = \frac{q_1(1 - q_1)n_1}{q(1 - q)n},$$

where $q_1 = n_A(t_1)/n(t_1)$ and $q = n_A/n$. This is the information fraction from the usual t-test. Then the spending function can be computed with respect to the $\tilde{t}$ values. See Lan and Lachin (1990) for more details on the appropriateness of using surrogate measures of information when the true information fraction cannot be computed.

For sequentially monitored conditional randomization tests, one must compute the proportion of the variance of the test statistics conditional on the number of patients assigned to treatment A. Let

$$t_l = \frac{a'_{n_1} \Sigma_{n_1|n_{1A}} a_{n_1}}{a'_n \Sigma_{n|n_A} a_n},$$

where $\Sigma_{n_1|n_{1A}} = \mathrm{Var}(\mathbf{T}|N_A(n_1) = n_{1A})$ and $\Sigma_{n|n_A} = \mathrm{Var}(\mathbf{T}|N_A(n) = n_A)$. Plamadeala and Rosenberger (2012) compute these conditional variance–covariance matrices exactly for Efron's BCD.

6.15 PROBLEMS

6.1 Verify equations (6.1) and (6.4).

6.2 Read Basu (1980) and the ensuing discussion (including that of the venerable discussant, the late Oscar Kempthorne). Prepare a 5 minute position paper expressing your views on randomization tests. Present your paper in a classroom debate with fellow students. Focus on the following issues:

(i) Did Fisher contradict himself or change positions on the role of randomization tests?
(ii) Does Basu make a convincing argument in his example of the scientist and the statistician?
(iii) Is there a middle ground?

6.3 For the examples in Tables 6.1 and 6.2, compute the following:

a. the unconditional and conditional p-values if Efron's BCD had been used with $p = 2/3$;
b. the p-value if the TBD had been used (see Table 3.2);
c. the conditional p-value for the UD(0, 1) if the actual randomization sequence had been *ABAA* instead of *ABBA*.

6.4 Let r_{jn} be the simple rank scores. The *van der Waerden scores* are defined as

$$a_{jn} = \Phi^{-1}\left(\frac{r_{jn}}{n+1}\right)$$

Lehmann (1975, p. 97), where Φ is the standard normal distribution function. Recompute the example in Table 6.1 using the van der Waerden scores instead of simple rank scores.

6.5 Verify equation (6.12).

6.6 Derive the variance in (6.14).

6.7 Show that condition (6.17) does not hold when

(i) $a_{jn} = q^j, 0 < q \neq 1$;
(ii) $a_{jn} = 1/j$. (Hájek, 1969)

6.8 Show that condition (6.17) holds for the van der Waerden scores, defined in Problem 6.4, at a rate $O(\ln n/n)$.
(*Hints*: (a) Use the approximation to the normal distribution function $1 - \Phi(x) \sim \phi(x)/x$ as $x \to \infty$, where $\phi(x)$ is the normal density function (e.g., Feller (1968, p. 175)); (b) Use the approximation $\bar{a}_n \sim \int_0^1 \Phi^{-1}(u)du$.)

6.9 Prove or give a counterexample to the following statement: "Any continuous function of the simple ranks satisfies (6.17)."

6.10 Redo the data analysis in Table 6.3 using the linear rank test with van der Waerden scores, defined in Problem 6.4, for cholesterol values in Table 6.6, assuming that Wei's UD(0, 1) was used to generate the treatment assignments.

6.11 The results of the randomization test can differ if a different re-randomization procedure is used. Compute an unconditional Monte Carlo re-randomization p-value with 20,000 replications, assuming that the linear rank test with simple centered ranks is used, under (i) Efron's BCD ($p = 2/3$) and (ii) complete randomization. The observed treatment assignments are $(1, 0, 0, 0, 1, 1, 1, 1, 1, 1)$ and the observed responses are $(1, 2, 3, 4, 5, 6, 7, 8, 9, 10)$.

6.12 Verify (6.26).

6.16 REFERENCES

AGRESTI, A. (1990). *Categorical Data Analysis*. John Wiley & Sons, Inc., New York.
ALTMAN, D. G. and ROYSTON, J. P. (1988). The hidden effects of time. *Statistics in Medicine* **7** 629–637.

BASU, D. (1980). Randomization analysis of experimental data: the Fisher randomization test. *Journal of the American Statistical Association* **75** 575–595, with discussion.

BERGER, V. W. (2000). Pros and cons of randomization tests in clinical trials. *Statistics in Medicine* **19** 1319–1328.

FELLER, W. (1968). *An Introduction to Probability Theory and its Applications*. Vol. **I**. John Wiley & Sons, Inc., New York.

FISHER, R. A. (1935). *The Design of Experiments*. Oliver and Boyd, Edinburgh.

FLEMING, T. R. and HARRINGTON, D. P. (1991). *Counting Processes and Survival Analysis*. John Wiley & Sons, Inc., New York.

GAIL, M. H., TAN, W. Y., and PIANTADOSI, S. (1988). Tests for no treatment effect in randomized clinical trials. *Biometrika* **58** 403–417.

GALBETE, A., MOLER, J. A., and PLO, F. (2015). Randomization tests in recursive response-adaptive randomization procedures. *Statistics*, in press.

GALBETE, A. and ROSENBERGER, W. F. (2015). On the use of randomization tests following adaptive designs. *Journal of Biopharmaceutical Statistics*, in press.

GASTWIRTH, J. L. (1966). On robust procedures. *Journal of the American Statistical Association* **61** 929–948.

HÁJEK, J. (1969). *A Course in Nonparametric Statistics*. Holden-Day, San Francisco, CA.

HÁJEK, J. and ŠIDÁK, Z. (1967). *Theory of Rank Tests*. Academia, Prague.

HOLLANDER, M. and PEÑA, E. (1988). Nonparametric tests under restricted treatment-assignment rules. *Journal of the American Statistical Association* **83** 1144–1151.

HU, F. and ROSENBERGER, W. F. (2006). *The Theory of Response-Adaptive Randomization in Clinical Trials*. John Wiley & Sons, Inc., New York.

JENNISON, C. and TURNBULL, B. W. (2000). *Group Sequential Methods with Applications to Clinical Trials*. Chapman and Hall/CRC, Boca Raton, FL.

KADANE, J. B. and SEIDENFELD, T. (1990). Randomization in a Bayesian perspective. *Journal of Statistical Planning and Inference* **25** 329–345.

KALBFLEISCH, J. D. and PRENTICE, R. L. (1980). *The Statistical Analysis of Failure Time Data*. John Wiley & Sons, Inc., New York.

KENNES, L. N., HILGERS, R.-D., and HEUSSEN, N. (2012). Choice of the reference set in a randomization test based on linear ranks in the presence of missing values. *Communications in Statistics–Simulation and Computation* **41** 1051–1061.

LACHIN, J. M. (1988). Statistical properties of randomization in clinical trials. *Controlled Clinical Trials* **9** 289–311.

LACHIN, J. M. (1999). Worst rank score analysis with informatively missing observations in clinical trials. *Controlled Clinical Trials* **20** 408–422.

LACHIN, J. M. (2000). *Biostatistical Methods: The Assessment of Relative Risks*. John Wiley & Son, Inc., New York.

LAN, K. K. G. and DEMETS, D. L. (1983). Discrete sequential boundaries for clinical trial. *Biometrika* **70** 659–663.

LAN, K. K. G. and LACHIN, J. M. (1990). Implementation of group sequential logrank tests in a maximum duration trial. *Biometrics* **46** 759–770.

LEHMANN, E. L. (1986). *Nonparametrics: Statistical Methods Based on Ranks*. Holden-Day, San Francisco, CA.

LIN, D. Y., WEI, L. J., and DEMETS, D. L. (1991). Exact statistical inference for group sequential trials. *Biometrics* **47** 1399–1408.

LITTLE, R. J. A. and RUBIN, D. B. (1987). *Statistical Analysis with Missing Data*. John Wiley & Sons, Inc., New York.

MATTS, J. P. and LACHIN, J. M. (1988). Properties of permuted block randomization in clinical trials. *Controlled Clinical Trials* **9** 345–364.

MCCULLAGH, P. and NELDER, J. A. (1989). *Generalized Linear Models*. Chapman and Hall, London.

MEHTA, C. R., PATEL, N. R., and WEI, L. J. (1988). Constructing exact significance tests with restricted randomization rules. *Biometrika* **75** 295–302.

PARHAT, P., ROSENBERGER, W. F., and DIAO, G. (2014). Conditional Monte Carlo randomization tests for regression models. *Statistics in Medicine* **33** 3078–3088.

PETO, R. and PETO, J. (1972). Asymptotically efficient rank invariant test procedures. *Journal of the Royal Statistical Society A* **135** 185–206, with discussion.

PLAMADEALA, V. and ROSENBERGER, W. F. (2012). Sequential monitoring with conditional randomization tests. *Annals of Statistics* **40** 30–44.

PRENTICE, R. L. (1978). Linear rank tests with right-censored data. *Biometrika* **65** 167–179.

ROSENKRANZ, G. K. (2011). The impact of randomization on the analysis of clinical trials. *Statistics in Medicine* **30** 3475–3487.

RUBIN, D. B. (1978). Bayesian inference for causal effects: the role of randomization. *Annals of Statistics* **6** 34–58.

SMYTHE, R. T. (1988). Conditional inference for restricted randomization designs. *Annals of Statistics* **16** 1155–1161.

SMYTHE, R. T. and WEI, L. J. (1983). Significance tests with restricted randomization design. *Biometrika* **70** 496–500.

WEI, L. J. and LACHIN, J. M. (1988). Properties of urn randomization in clinical trials. *Controlled Clinical Trials* **9** 345–364.

ZHANG, L. and ROSENBERGER, W. F. (2011). Adaptive randomization in clinical trials. In *Design and Analysis of Experiments: Special Designs and Applications*, Vol. **3** (HINKELMANN, K., ed.). John Wiley & Sons, Inc., New York, Chapter 7.

6.17 APPENDIX A

Table 6.6 gives data from the Diabetes Control and Complications Trial used in the data analysis example in Section 6.8.2.

Table 6.6 *Cholesterol levels and treatment assignment codes from 50 patients from the Diabetes Complications and Control Trial.*

Patient	Cholesterol	Treatment Assignment
1	132	1
2	195	0
3	157	0
4	196	0
5	190	1
6	228	1
7	191	1
8	150	1
9	154	0
10	147	1

(continued)

Table 6.6   (*Continued*)

Patient	Cholesterol	Treatment Assignment
11	207	0
12	113	1
13	174	0
14	210	1
15	144	0
16	217	0
17	167	1
18	229	1
19	123	1
20	248	1
21	146	0
22	193	0
23	182	0
24	116	0
25	189	0
26	215	1
27	211	1
28	252	0
29	206	1
30	151	1
31	232	0
32	238	1
33	201	0
34	174	1
35	151	1
36	150	1
37	221	0
38	232	1
39	179	1
40	167	0
41	213	0
42	153	0
43	122	0
44	204	1
45	196	1
46	168	1
47	196	1
48	185	1
49	235	1
50	143	0

6.18   APPENDIX B

Here we present the SAS code to compute a Monte Carlo unconditional randomization test, based on the linear rank test with simple rank scores, under the UD(0, 1)
randomization procedure. The user specifies the number of re-randomizations and the
seed for the random number generation. The resulting p-value is given in Table 6.3.

```
data dcct;
 input id chol tx;
cards;
1 132 1
 .
 .
 .
;
*First, the centered ranks are computed;
proc rank out=ranked;
 ranks simrank;
 var chol;
run;
proc means data=ranked noprint;
 var simrank;
 output out=b mean=meanrank n=numobs;
run;
data two;
 set ranked;
 if _N_=1 then set b;
 cenrank=simrank-meanrank;
 keep id cenrank tx numobs;
*The observed test statistic is computed;
data three;
 set two;
 na+tx;
 if tx=1 then ranktest+cenrank;
 if id=50 then sobs=ranktest;
*The centered ranks are transposed so they can be read
 into an array;
proc transpose data=three name=name prefix=cenrank
 out=four;
 var cenrank;
run;
*A data set is created with the observed test statistic
 and the sample size, to be merged with the centered
 ranks;
data five;
 set three;
 if id=50;
 name='cenrank';
 keep sobs name numobs;
*The re-randomization is performed l times and the test
 statistic is recomputed each time;
*A counter is incremented every time the test statistic
 exceeds the observed test statistic;
```

```
*The user specifies l and the random number generator
 seed;
data six;
 merge four five;
 by name;
 l=;
 seed=;
 array y[50] cenrank1-cenrank50;
 pcount=0;
 do i=1 to l;
  sumrank=0; na=0; nb=0;
  do j=1 to numobs;
   cenrank=y[j];
   if j=1 then p=1/2;
   else p=nb/(j-1);
   x=ranuni(seed);
   if x<p then newt=1;
   else newt=0;
   na+newt;
   nb=j-na;
   if newt=1 then sumrank+cenrank;
   if j=numobs then do;
    s=sumrank;
    if abs(s)>abs(sobs) then pcount+1;
   end;
  end;
 end;
pvalue=pcount/l;
proc print;
 var pvalue;
run;
```

7

Stratification

7.1 INTRODUCTION

Chapter 1 pointed out the likelihood of confounders in biomedical studies. In any clinical trial, there are *covariates* or *prognostic factors* of interest besides the treatment effect. Some covariates are known in advance to be important risk factors that are associated with the outcome of a patient. For instance, in trials of heart disease, relevant covariates may be cholesterol, blood pressure, age, or gender. In a randomized study, one objective is to equalize the distribution of such factors within each treatment group so as to minimize biases due to heterogeneity. The most common covariate causing such concern in multicenter clinical trials is the clinical center, since clinics usually differ with respect to the demographic and clinical characteristics of their patient populations and adherence to the protocol and various procedures. Thus, an imbalance in the numbers randomized to each group within a clinic, such as 60% to A in clinic 1 and 30% to A in clinic 2, may bias the results of the study. As we have seen in Chapter 4, randomization tends to equalize the distribution of both known and unknown covariates across the treatment groups. However, known covariates that can introduce heterogeneity into the clinical trials population are often of special concern to the investigator.

Although randomization tends to mitigate the possibility of serious covariate imbalances among the treatment groups, it is not unusual for imbalances to occur in practice. In this event, stratified analyses or regression modeling can be used to adjust for important covariates in a *post-hoc* analysis. Another alternative is to implement a design that ensures balance on specified covariates in the trial. In this chapter, we discuss randomization techniques that tend to balance treatments within the discrete levels, or *strata*, of known covariates. We describe the relative merits of stratified randomization.

Randomization in Clinical Trials: Theory and Practice, Second Edition.
William F. Rosenberger and John M. Lachin.
© 2016 John Wiley & Sons, Inc. Published 2016 by John Wiley & Sons, Inc.

7.2   STRATIFIED RANDOMIZATION

In a randomized study, *stratification* has interchangeably been used to refer to either a stratified randomization (often termed *prestratification*), or a stratified-adjusted analysis (often termed *poststratification*) with or without a stratified randomization. To avoid ambiguity, the distinction is drawn between a stratified randomization and a stratified-adjusted analysis, each as described as follows.

In a stratified randomization (prestratification), subjects are grouped according to covariate values prior to randomization, and subjects are then randomized within strata. Within each stratum, a separate randomization sequence is employed. For example, consider a study stratified by clinic (say 5 in number) and gender, with 10 total strata defined jointly by the covariates clinic with 5 categories and gender with 2. In this case, a separate randomization sequence would be employed for each gender and for each clinic, 10 sequences in all. A female subject recruited by clinic 3 would be randomized using the "clinic 3 and female" randomization sequence. If a restricted randomization sequence is employed, then the probability of assignment of this subject would depend only on the number of prior assignments to A and B among females recruited by clinic 3, and not on the prior treatment assignments to males recruited by clinic 3, nor on the prior assignments to males or females recruited by other clinics. In this case, the randomization sequences within the 10 strata would be accomplished by using 10 separate restricted randomization sequences.

Stratification is also used to refer to postrandomization stratification in the analysis whereby the subjects are grouped within strata or subgroups according to one or more patient characteristics. In the aforementioned example, the analysis might be performed using the 10 strata defined on the joint basis of the covariates "clinic" and "gender." The first stage of the analysis is to compare treatments A versus B separately within each strata. This is also called a *subgroup analysis*. In the second stage, various methods might then be used to perform a combined test, by pooling the results of all the strata or subgroups in some way, so as to provide an aggregate test over strata or subgroups.

For the aforementioned example of a randomization stratified by clinic and gender, consider the simplest case of two treatment groups A versus B and a binary response (e.g., "healed" versus "not healed"). Within each clinic–gender stratum, a 2×2 table can be formed to assess the treatment–response association within that stratum, expressed as an odds ratio. To then assess the aggregate treatment–response association over all 10 strata combined, the Mantel–Haenszel procedure could be applied. This provides a stratified-adjusted estimate of the common odds ratio, obtained as a linear combination of the within-strata odds ratios; and an aggregate stratified-adjusted test of association. The analysis effectively adjusts for the stratum effects since treatments A and B are compared within strata and then the results are combined over strata (cf. Lachin, 2010).

An alternate strategy is to simply combine patients and responses over all strata into a single 2×2 table for which A and B are then compared in a single analysis, ignoring any strata, even those used as a basis for a stratified randomization. In this case, the stratified randomization is effectively ignored in the analysis. This *pooled analysis* is also called a *combined*, *unstratified*, or *unadjusted analysis*.

Note that a stratified-adjusted analysis can be performed for any groupings of subjects regardless of whether the randomization was stratified according to those groupings. Conversely, an unstratified analysis can be performed, even though the randomization may have been stratified by other factors. Thus, an initial consideration might be the relative efficiency of a stratified-adjusted analysis with a stratified versus nonstratified randomization and the relative efficiency of a stratified versus unstratified analysis of a study that employed stratified randomization. We will explore these issues in detail.

7.3 IS STRATIFICATION NECESSARY?

The gains from stratification were early recognized in sample surveys where it was shown that a stratified analysis improves the precision of estimators. However, it is principally the stratified analysis that eliminates bias, for which a stratified randomization is not necessary. Thus, it should not be surprising that a stratified randomization tends to improve the efficiency of estimators and power of tests in a small trial, say for $n < 100$, but has negligible advantage in large trials. However, in clinical trials where there is a large difference in the size of strata, this advantage tends to disappear (Ganju and Zhou, 2011).

Stratification is often considered to be an essential feature of randomization (cf. Zelen, 1974), but there has been significant controversy as to the relative statistical merits of stratified randomization versus a stratified analysis following unstratified randomization. Permutt (2007) discusses three problems in clinical trials in relation to stratification. The first is a treatment-by-covariate interaction. In this case, a treatment may have a differential response in different subgroups of patients. Permutt indicates that a stratified *analysis* is important to identify heterogeneity among subgroups, but that stratified randomization is less important than is often believed:

> It would indeed be disastrous if all the elderly patients, for example, were assigned to one treatment because there would then be no comparison to make within the stratum of elderly patients. Randomization, however, whether or not stratified, makes this outcome vanishingly improbable unless there are very few elderly patients. The case of a very small stratum, however, raises a much more important problem.... It might seem desirable to maximize the efficiency within the stratum by ensuring that equal numbers of elderly patients were assigned to each treatment, but the advantage is slight. If the elderly patients happen to be allocated 15-35 and the younger patients 35-15, the standard error... [is] only slightly different: and an imbalance even this large happens only about one time in 8,000 with unstratified allocation. It is sometimes incorrectly supposed that stratified allocation is necessary to draw valid inferences from the separate analyses. Even without stratification, however, the patients within a stratum are allocated according to a sequence of random numbers. It is of no consequence that these random numbers happen to be drawn from a larger set, skipping some that were used in other strata.

The second problem is the concern about interactions in small strata. In this case, Permutt indicates that the solution is *stratified recruitment* to ensure adequate representation of subgroups.

The third problem is confounding due to a main effect of stratum. Here the concern is that a covariate imbalance between treatment groups may confuse the issue of whether there is a beneficial treatment effect. However, these covariates can be adjusted for in a stratified analysis, and such an analysis provides valid inference whether or not the covariate is balanced across treatment groups.

The latter point has been misunderstood by investigators and journal reviewers (Senn, 1994, 2013). In fact, every report of clinical trials in the literature includes a table of baseline covariates, sometimes with associated *p*-values comparing the two treatment groups. While randomization always works as intended, some of these covariates will be unbalanced simply due to making a type I error—in fact, 5 percent of the covariates should be unbalanced due to chance alone, if they are independent. A nonsignificant *p*-value should not give any comfort, as failure to reject the null hypothesis does not prove the null hypothesis. This "fear" of covariate imbalances leads some investigators to propose overstratified trials with very small numbers in some strata. Senn puts it this way:

> I think we should avoid pandering to these foibles of physicians .... I think people worry far too much about imbalance from the inferrential (sic) point of view.... The way I usually describe it to physicians is as follows: if we have an unbalanced trial, you can usually show them that by throwing away some patients you can reduce it to a perfectly balanced trial. So you can actually show that within it there is a perfectly balanced trial. You can then say to them: 'now, are you prepared to make an inference on this balanced subset within the trial?' and they nearly always say 'yes'. And then I say to them, 'well how can a little bit more information be worse than having just this balance trial within it?' [from the discussion on Atkinson (1999)]

Thus, it is often recommended that the randomization for a clinical trial should be stratified only on those factors considered absolutely necessary to ensure the integrity of the study (cf. Friedman, Furberg, and DeMets, 1998; Peto *et al.*, 1976). In many studies, differences among clinics are the major source of heterogeneity in the outcome measures. Further, since a clinic frequently may withdraw (or be dropped) from a study, it is desirable that such withdrawal should not affect the validity of the overall randomization plan. For these reasons, it is also generally advocated that randomization in a multicenter trial should be stratified by clinic.

7.4   TREATMENT IMBALANCES IN STRATIFIED TRIALS

A common misconception is that stratified randomization promotes greater balance between the numbers of treatment assignments to A and B within each stratum and thus overall. Unfortunately, this is not always so.

Since some randomization procedures, including complete randomization, Efron's biased coin design, and Wei's urn, do not force balance between treatments (except asymptotically), there is a positive probability that imbalances will occur within individual strata when stratified randomization is performed. Let N_{iA} be the number of patients assigned to treatment A in stratum $i, i = 1, \ldots, s$, then $N_A(n) = \sum_{i=1}^{s} N_{iA}$. Asymptotically, $N_A(n)$ should be approximately $n/2$. However, for finite samples, with a large number of small strata, imbalances are additive across strata and can result in an overall imbalance of some significance. This is less likely to occur when there are small numbers of large strata.

Often a permuted block design is used within each stratum to ensure balance. This is called *stratified blocked randomization*. With the random allocation rule or permuted block design, there is no imbalance within strata or in aggregate as long as all blocks are filled. However, if some blocks are not filled, a treatment imbalance can occur. Since an unstratified randomization risks at most one unfilled block, whereas a s-strata randomization risks at most s unfilled blocks, the probability of a treatment imbalance is greater in a clinical trial with stratified randomization.

In stratified blocked randomization, one must be careful to limit the stratification variables and the number of strata within each to a minimum, representing only the most important variables and levels. For instance, in a multicenter trial with 15 participating institutions, stratifying by clinic, gender, race (3 levels), and age (2 levels) leads to $15 \times 2 \times 3 \times 2 = 180$ strata. Unless the trial is extremely large, some strata will have very few patients.

Hallstrom and Davis (1988) describe the probability of aggregate imbalances in a trial when using stratified blocked randomization. Suppose n patients are assigned with equal probability to treatments A and B, within s strata. Each of the s strata is balanced by using permuted blocks, and the block size in the ith stratum is $m_i, i = 1, \ldots, s$. Let N_i be the number of patients assigned in the last block of stratum i and A_i be the number assigned to treatment A, $1 \leq A_i \leq N_i \leq m_i$. Define $D_i = N_i - 2A_i$ to be the imbalance in the ith stratum. Conditional on N_i, A_i has a hypergeometric distribution, with expectation $N_i/2$ and variance $N_i(m_i - N_i)/4(m_i - 1)$. Then $E(D_i|N_i) = 0$ and $E(D_i) = 0$. We can then derive

$$\text{Var}(D_i) = \frac{E\{N_i(m_i - N_i)\}}{m_i - 1} \tag{7.1}$$

(see Problem 7.2). Summing over independent strata, the total imbalance in the trial is given by

$$D = \sum_{i=1}^{s} D_i.$$

Using equation (7.1), we can determine the effect of block size and the number of strata on the variability of D, provided we have some information on the distribution of N_i. Hallstrom and Davis (1988) consider two models. The first model assumes that the expected number of patients in each stratum is large relative to the block size. In this case, it is reasonable to assume that N_i follows a discrete uniform distribution on the support $\{1, \ldots, m_i\}$. Then $E(N_i) = (m_i + 1)/2$ and $E(N_i^2) = (m_i + 1)(2m_i + 1)/6$.

From (7.1), we can compute $\text{Var}(D_i) = (m_i + 1)/6$. So, under the uniform model, we have

$$\text{Var}(D) = \frac{\sum\limits_{i=1}^{s} m_i + s}{6}. \tag{7.2}$$

The normal approximation can be used to compute

$$\Pr(|D| > d) \cong 2\left\{1 - \Phi\left(\frac{d}{\sqrt{\text{Var}(D)}}\right)\right\} \tag{7.3}$$

for various values of m_i. Such an exercise is useful in planning studies.

For example (see Hallstrom and Davis (1988, p. 378)), the Cardiac Arrhythmia Suppression Trial (CAST) was planned with a total of 4200 patients and 270 strata. Using equation (7.2) with $m_i = 4$, we find that $\text{Var}(D) = 225$. Then by equation (7.3), we can compute $\Pr(|D| > 30) = 0.05$. Such a difference is small compared to the number of patients and would not be of concern. The maximum imbalance is $\sum_{i=1}^{s} m_i/2 = 540$.

Anisimov (2010) computes the distribution of an imbalance in stratified randomization by incorporating a model on patient recruitment, predicting the number of patients randomized in each stratum. He concludes that the imbalance induced by stratifying on clinical center increases the sample size by only a few patients, compared to unstratified randomization.

7.5   STRATIFIED ANALYSIS USING RANDOMIZATION TESTS

The importance of conducting a stratified test following a stratified randomization has been well established under a population model (e.g., Kahan and Morris, 2011), because the unadjusted standard errors of the treatment effect estimators will be inflated due to within-stratum correlation. Nonetheless, many clinical trials do not conduct stratified analyses following stratified randomization (Kahan and Morris, 2012).

Under a randomization model, the proper analysis for any clinical trial with a stratified randomization is a *like-stratified analysis*. In a like-stratified analysis, patients within the same stratum are compared, and a stratified test is computed by summing the stratum-specific tests over strata. Because strata may have differential sample sizes, often a weighted test is used.

The procedure using Monte Carlo re-randomization techniques is clear from Section 6.9. Within each stratum $i = 1, \ldots, I$, an observed test statistics $S_{\text{obs.},i}$ is computed. The stratified observed test is computed as

$$S_{\text{obs.}} = \sum_{i=1}^{I} \omega_i S_{\text{obs.},i}, \tag{7.4}$$

where $\omega_i \in [0,1], i = 1, \dots, I$ are weights such that $\sum_{i=1}^{I} \omega_i = 1$. Within each stratum, the randomization sequence is regenerated L times, and a new stratum-specific test statistic $S_{il}, i = 1, \dots, I, l = 1, \dots, L$ is computed. We then have L Monte Carlo stratified tests, computed as $S_i = \sum_{i=1}^{I} \omega_i S_{il}$, and the stratified p-value estimator is computed using (6.22). Conditional tests, conditioning on the numbers of patients assigned to treatments A and B in each stratum, can be generated similarly using techniques described in Section 6.9.3.

There are a number of subtle issues regarding the implementation of these stratified randomization tests in practice. First, the stratum-specific weights must be specified. In general, one would wish to weight strata by their relative importance in the analysis of the clinical trial. From a population model perspective, optimal weights can be derived according to efficiency considerations under certain forms of the alternative hypothesis (e.g., van Elteren, 1960; Puri, 1965). However, under a randomization model, such optimality criteria do not apply. Weighting by the stratum-specific sample fractions, that is, $\omega_i = n_i/n$, where n_i is the total number of patients in stratum $i = 1, \dots, I$, is a simple technique that is probably the most sensible in practice.

In most trials, the randomization is stratified by clinical center, and often there is a large disparity in the size of the strata, with some clinics recruiting a small number of subjects. In many instances, these small strata are pooled to form one strata of size comparable to that of other clinics. However, there is no randomization basis for doing so. Further, small strata still contribute to the test of the group difference, provided that at least one subject in each stratum is assigned to each group. In the case that no subjects are assigned to a group within a stratum, then that stratum cannot contribute to a comparison of treatments under a randomization model and would be discarded.

The necessity of discarding strata with only one patient in a stratified randomization test is controversial. Flyer (1998) describes several stratified trials where 6 to 8 percent of the data would have to be discarded. An unstratified analysis could be performed on the complete data set by computing an unstratified test across all patients. While the re-randomization would be stratum-specific, the analysis would not be stratified. In this way, patients alone in a stratum would still be included in the reference set. Flyer (1998) addresses this issue by comparing the efficiency of randomization tests that discard data to an analysis that does not condition on the stratum size. The determining factor is whether the between-stratum variance outweighs the information contained in the discarded strata. As a simple extreme example, suppose every stratum has one patient. Within each stratum, regardless of the restricted randomization procedure used, that lone patient was assigned to treatment with probability $1/2$, so across independent strata, complete randomization was used. A stratified analysis could not be conducted; however, across strata, a randomization test under complete randomization could be computed.

A stratified analysis is one approach to adjust for any bias introduced by an uncontrolled covariate, or to increase the efficiency by accounting for a highly influential covariate, whether or not the randomization was stratified on that covariate. This approach, however, is only applicable to qualitative covariates, or discretized quantitative covariates, and few in number. In many respects, it is more natural to perform an adjustment using a regression model that allows for both qualitative and quantitative covariates simultaneously. The techniques for doing under a randomization model were described in Section 6.11.

In addition to the overall comparison of treatments A and B, it may also be desired to compare treatments separately among those patients who are members of a subgroup defined *post hoc* on the basis of a covariate, usually a baseline (prerandomization) characteristic not used as a basis for stratification in the randomization. Such poststratified analyses are used to obtain a stratified-adjusted assessment of the overall treatment effect. For example, the treatments might be compared separately among men and separately among women in a study where the randomization was not stratified by gender. A covariate-adjusted test of treatment effect might then be obtained by combining the separate tests for men and women.

While such stratified analyses may in fact be specified *a priori* (in fact, such specification is preferred), such analyses are *post hoc* with respect to the generation of the randomization. Any such analysis specified after examination of the data could be criticized unless the basis for the analysis was specified *a priori*. For example, the protocol might specify that a *post hoc* stratified analysis would be conducted to adjust for any covariates on which the groups differed significantly by chance.

In order to perform a valid analysis among the subsets of patients within such subgroups, it is sufficient to assume that the randomly assigned treatment indicator variable values $\{T_j\}$ are statistically independent of the covariate values $\{X_j\}$ among the n patients randomized. We refer to this as the *covariate independence assumption*, which specifically assumes that $E(T_j|\mathcal{F}_{j-1}, X_j) = E(T_j|\mathcal{F}_{j-1})$ where $\mathcal{F}_{j-1}$ is the history of prior allocations, as would apply to a restricted randomization procedure. Clearly, this assumption is satisfied for any baseline covariate when there is no potential for selection bias.

For complete randomization, since the probability of treatment assignment to A is $E(T_j) = 1/2$ independently for all patients, the randomization test can be performed within any subgroup as though a separate randomization had been performed within that subgroup. Further, the test statistics within each of multiple strata will be statistically independent, and thus can be combined, exactly as though stratified randomization had been performed. For any other randomization procedure, since the probabilities of assignment are not independent and identically distributed, the validity of such analyses rests on the covariate independence assumption. For example, this assumption could be violated if the randomization is open to selection bias.

7.6 EFFICIENCY OF STRATIFIED RANDOMIZATION IN A STRATIFIED ANALYSIS

One question that often arises in the planning of a clinical trial is whether a stratified randomization will increase the precision of a stratified analysis. The relative efficiency of a stratified test with stratified randomization versus without stratified randomization was assessed by Grizzle (1982) using a homogeneous population model and by Matts and McHugh (1978) using a randomization model. Essentially identical results were obtained.

Following Grizzle's treatment, assume that the subjects arise from a homogeneous population over time within each stratum. For example, if patient gender is the covariate of interest, whether in a study with randomization stratified by gender or one unstratified, we assume that the likelihood that a male will enter the study is a constant over time. This assumption is equivalent to assuming that the covariate values are independently and identically distributed within each stratum.

Grizzle assessed this issue as follows. Consider the simplest case of two treatments ($k = A$ or B) and two strata ($i = 1, 2$) in a simple additive linear model

$$Y_{ijk} = \alpha + \beta_i + \mu_k + \epsilon_{ijk}, \tag{7.5}$$

for $j = 1, \ldots, n_{ik}$ subjects on treatment k in stratum i. As usual, the errors are assumed to be independent and identically distributed with $E(\epsilon_{ijk}) = 0$ and $\text{Var}(\epsilon_{ijk}) = \sigma^2$. The effect of the kth treatment is μ_k and that of the ith stratum is β_i, with $\beta_1 + \beta_2 = 0$. The treatment effect is $\theta = \mu_1 - \mu_2$. It is important to note that there is no treatment-stratum interaction. Thus, treatment effects within strata are assumed to be homogeneous, thus maximizing the gains in power from a stratified analysis.

In the case of a stratified randomization, it is assumed that the sample sizes allocated to each treatment are always equal, either in total or within strata. That is, we assume that the stratified randomization was 100 percent effective in eliminating covariate imbalances on the stratifying covariate(s). This is guaranteed if a random allocation rule or permuted block randomization is employed (with all blocks filled) and is the expectation with the other randomization procedures (complete or urn randomization).

Now consider the case of an unstratified randomization. Let $q_k = n_{1k}/n_k, k = A, B$, be the proportion of subjects randomized to the kth group who are also members of the first stratum ($i = 1$), and $1 - q_k$ be the proportion of those in the kth group who are members of the second stratum ($i = 2$). A covariate imbalance occurs when $q_A \neq q_B$. Conversely, with 100 percent-effective stratified randomization, it is assumed that the covariate stratum fractions are fixed and equal, $q_A = q_B$, so that there is no covariate imbalance.

Now consider the efficiency of an estimator of θ. Denote the variance of the estimator with stratified randomization (r) and stratified analysis (a), such that $q_A = q_B$, as $\sigma^2_{\theta(r,a)}$. Likewise, denote the variance of the estimator with unstratified randomization but with a subgroup analysis as $\sigma^2_{\theta(a)}$. When $n_A = n_B$, the relative efficiency of the estimators is then

$$R.E. = \frac{\sigma^2_{\theta(r,a)}}{\sigma^2_{\theta(a)}}.$$

Using the least squares estimator from (7.5), Grizzle (1982) shows that

$$\sigma^2_{\theta(a)} = \frac{4\sigma^2}{n} \left[\frac{q_A + q_B}{q_A(1 - q_A) + q_B(1 - q_B)} \right] \left[1 - \frac{q_A + q_B}{2} \right]. \tag{7.6}$$

Then with stratified randomization, taking $q_A = q_B$ in (7.6), we obtain

$$\sigma^2_{\hat{\theta}(r,a)} = \frac{4\sigma^2}{n}.$$

The relative efficiency is then given by

$$R.E. = \left\{ \frac{q_A + q_B}{q_A(1 - q_A) + q_B(1 - q_B)} \left[1 - \frac{q_A + q_B}{2} \right] \right\}^{-1}. \tag{7.7}$$

Note that $R.E. \le 1$ and $R.E. = 1$ when $q_A = q_B$. This relative efficiency of the estimators is also proportional to the relative power of a statistical test of $H_0 : \theta = 0$ using a *post hoc* stratified analysis versus a stratified randomization and a stratified analysis. Grizzle (1982) also gives the relative error for the case where $n_A \ne n_B$. From equation (7.7), Table 7.1 can be computed, which gives the relative efficiencies for various values of q_A and q_B ranging from 0.10 to 0.90. Lachin (2000, Section 3.5.4) presents an equivalent model for the analysis of binary data.

Now, suppose that q_A and q_B are binomial random variables with $E(q_A) = E(q_B) = \gamma$. For a given value γ, we can use the normal approximation to compute the probability that their absolute difference exceeds some value r. This is given by

$$\Pr(|q_A - q_B| > r) = 2 \left\{ 1 - \Phi \left(\frac{r\sqrt{n}}{2\sqrt{\gamma(1 - \gamma)}} \right) \right\}.$$

For various values γ ranging from 0.1 to 0.9, and for various sample sizes n, Table 7.2 gives the limits of imbalance, which would occur with probabilities 0.05 and 0.01. These imbalances can then be used with Table 7.1 to assess the loss of efficiency due to nonstratification.

For example, for $n = 25$ and $\gamma = 0.5$, an imbalance of 0.7 and 0.3, respectively, could occur with $p \cong 0.05$, which would result in an efficiency of 0.84 with

Table 7.1   *Relative efficiency of estimators for stratified randomization and stratified analysis versus stratified analysis only, for various values of q_A and q_B.*

q_A	q_B	$R.E.$
0.3	0.1	0.938
0.5	0.1	0.857
0.5	0.3	0.821
0.7	0.1	0.625
0.7	0.3	0.840
0.7	0.5	0.958
0.9	0.1	0.360
0.9	0.3	0.625
0.9	0.5	0.810
0.9	0.7	0.938

Table 7.2 *Limits of imbalance occurring with probabilities 0.05 and 0.01, for various values of n and γ.*

n	γ	0.05	0.01
25	0.1	0.235	0.309
25	0.3	0.359	0.472
25	0.5	0.392	0.515
50	0.1	0.166	0.219
50	0.3	0.254	0.334
50	0.5	0.277	0.364
100	0.1	0.116	0.154
100	0.3	0.180	0.236
100	0.5	0.196	0.258
200	0.1	0.083	0.109
200	0.3	0.127	0.167
200	0.5	0.139	0.182

unstratified randomization. However, for $n = 100$, there is probability < 0.01 of covariate Imbalances, which would result in a relative efficiency of 0.9 or less.

The aforementioned results apply regardless of the method of randomization or treatment assignment employed because a homogeneous population model is assumed. A randomization-based analysis of this same issue was explored by Matts and McHugh (1978) assuming that a random allocation rule was employed with and without stratification. Again, note that the random allocation rule guarantees equal sample sizes within each group in total and within each stratum. Matts and McHugh also use a simple linear model like (7.5), but allow for more than two treatment groups and an arbitrary number of strata.

For the case of only two equal sized groups, they show that the relative efficiency for a study of size n, with s strata of equal size $n_i = n/s$, is obtained as

$$R.E. = \frac{n\left[1 - \left(1 - \frac{1}{s}\right)^n\right]}{n + 2s - 2}. \tag{7.8}$$

Clearly, as n increases relative to s, $R.E. \to 1$. Solving for n in (7.8) and ignoring the asymptotically negligible term $\{(s-1)/s\}^n$, we obtain

$$n \cong \frac{2R.E.(s-1)}{1 - R.E.}.$$

For example, for $s = 10$, a stratified analysis with unstratified randomization will yield 90 percent of the efficiency of a stratified randomization with a stratified analysis for a sample size of $n \cong 162$ or greater. For $s = 10$, 95 percent efficiency is provided by $n \cong 342$ or greater.

Therefore, under two entirely different approaches, it has been shown that a stratified randomization will have nonnegligible effects on the efficiency of a stratified analysis with small sample sizes, but that as the sample size increases, there are miniscule gains in efficiency from a stratified randomization relative to a simple *post hoc* stratified analysis. Unfortunately, even though stratification has greatest merit in small trials, it is usually not feasible to stratify on more than one or two factors due to the small within-stratum cell sizes.

7.7 CONCLUSIONS

For a randomization-based analysis, prestratification has several advantages. First, one can discard strata based on *a priori* operational criteria without affecting the randomization stream. This is particularly relevant for prestratification by clinic in multicenter clinical trials, where a clinic may later be discarded in the analysis, such as when a clinic's participation is terminated due to lack of recruitment. Second, prestratification allows for a very simple stratified analysis by simply summing the test statistic over the independent strata, possibly with stratum-specific weights. However, it should be clear from the developments in Section 7.6 that prestratification does not result in any benefits in terms of relative efficiency for a stratified analysis in large sample clinical trials.

In some cases, a poststratified analysis may be desired for covariates not considered in the prestratification process. One option is to conduct such an analysis with a separate test of treatment effects within strata. The within-stratum tests can then be combined to provide an overall stratified-adjusted assessment of the difference between the treatment groups.

If the objective of poststratification is to provide an adjusted assessment of treatment effect, a simpler approach is to conduct a randomization-based test in a modeling setting, as discussed in Section 6.11. This also has the advantage that a randomization-based test can be conducted after adjusting for multiple covariates, including quantitative covariates.

Every trial includes *confirmatory* analyses of primary, and often secondary, objectives. For such analyses, it is important that the principal method of analysis be specified *a priori*, whether randomization-based or population-based, whether adjusted for other covariates, and, if so, how the covariates are to be selected and how the adjustment is to be performed. However, after a trial has been conducted, a variety of *exploratory* analyses are conducted to address objectives beyond those explicitly stated in the protocol. Such analyses could be performed either under a randomization model or under an invoked population model, and many statisticians would favor a population model for covariate-adjusted regression analyses when considering such hypotheses.

7.8 PROBLEMS

7.1 Read Senn (1994) and Permutt (2007). Prepare a two-page position paper on whether and when stratification is necessary, incorporating thoughts presented in those papers as well as your own opinion.

7.2 Derive equation (7.1).

7.3 Refer to the CAST example in Section 7.4. Suppose each clinic contains four hospitals, so that the number of strata becomes $270 \times 4 = 1080$. Determine the probability than an imbalance greater than 30 will result and the maximum possible imbalance assuming that N_i is uniform. Comment on the appropriateness of stratified blocks in this setting (Hallstrom and Davis, 1988).

7.4 Simulate a stratified clinical trial with $I = 4$ strata. Assume that the big stick design was used ($b = 3$) within each stratum sample sizes $n_1 = 50, n_2 = 25, n_3 = 10$, and $n_4 = 50$. Assume that patient responses arise from the following i.i.d. population model within each stratum:

Stratum 1, 2: A: $N(0, 1)$; B: $N(1, 1)$;

Stratum 3, 4: A: $N(1, 1)$; B: $N(0, 1)$.

a. Compute $S_{\text{obs.}}$ using equation (7.4) for a single specific generated randomization sequence and response sequence. Compute the p-value for a stratified randomization test, holding the responses fixed, using weights $\omega_i = n_i/n$. Generate $L = 15,000$ new randomization sequences. Note that the treatment effect is reversed in small and large strata. How do the weights impact the analysis?
b. Redo the analysis using an unstratified randomization tests. How do the results compare?

7.9 REFERENCES

ANISIMOV, V. V. (2010). Effects of unstratified and centre-stratified randomization in multi-centre clinical trials. *Pharmaceutical Statistics* **10** 50–59.

ATKINSON, A. C. (1999). Optimum based-coin designs for sequential treatment allocation with covariate information. *Statistics in Medicine* **18** 1741–1752 (with discussion).

VAN ELTEREN, P. H. (1960). On the combination of independent two-sample tests of Wilcoxon. *Bulletin of the International Statistical Institute* **37** 351–361.

FLYER, P. A. (1998). A comparison of conditional and unconditional randomization tests for highly stratified designs. *Biometrics* **54** 1551–1559.

FRIEDMAN, L. M., FURBERG, C. D., AND DEMETS, D. L. (1998). *Fundamentals of Clinical Trials*. Springer-Verlag, New York.

GANJU, J. AND ZHOU, K. (2011). The benefit of stratification in clinical trials revisited. *Statistics in Medicine* **30** 2881–2889.

GRIZZLE, J. E. (1982). A note on stratifying versus complete random assignment in clinical trials. *Controlled Clinical Trials* **3** 365–368.

HALLSTROM, A. AND DAVIS, K. (1988). Imbalance in treatment assignments in stratified blocked randomization. *Controlled Clinical Trials* **9** 375–382.

KAHAN, B. C. AND MORRIS, T. P. (2011). Improper analysis of trials randomised using stratified blocks or minimisation. *Statistics in Medicine* **31** 328–340.

KAHAN, B. C. AND MORRIS, T. P. (2012). Reporting and analysis of trials using stratified randomisation in leading medical journals: review and reanalysis. *British Medical Journal* **345** 1–8.

LACHIN, J. M. (2010). *Biostatistical Methods: The Assessment of Relative Risks*. John Wiley & Sons, Inc., New York.

MATTS, J. P. AND McHUGH, R. B. (1978). Analysis of accrual randomized clinical trials with balanced groups in strata. *Journal of Chronic Diseases* **31** 725–740.

PERMUTT, T. (2007). A note on stratification in clinical trials. *Drug Information Journal* **41** 719–722.

PETO, R., PIKE, M. C., ARMITAGE, P., BRESLOW, N. E., COX, D. R., HOWARD, S. V., MANTEL, N., McPHERSON, K., PETO, J., AND SMITH, P. G. (1976). Design and analysis of randomized clinical trials requiring prolonged observation of each patient—I: introduction and design. *British Journal of Cancer* **34** 585–612.

PURI, M. L. (1965). On the combination of independent two-sample tests of a general class. *International Statistical Review* **33** 229–241.

SENN, S. (1994). Testing for baseline imbalance in clinical trials. *Statistics in Medicine* **13** 1715–1726.

SENN, S. (2013). Seven myths of randomisation in clinical trial. *Statistics in Medicine* **32** 1439–1450.

ZELEN, M. (1974). The randomization and stratification of patients to clinical trials. *Journal of Chronic Diseases* **28** 365–375.

8

Restricted Randomization in Practice

8.1 INTRODUCTION

Each method of randomization has properties that are better suited to specific applications than others. Thus, the choice of a randomization procedure and its implementation depend in part on the design features of the study. In this chapter, we outline the basic steps in determining an appropriate randomization procedure to use, generating the randomization sequence, and implementing the randomization in the clinical trial.

The paramount objective of randomization is to provide an unbiased comparison of the treatment groups. However, if not carefully implemented, the randomization procedure can be subverted, even in a double-masked study. Also, randomization procedures in an unmasked study are susceptible to varying degrees to subtle biases introduced by the investigators (selection bias) and to subtle biases introduced in any study by nonrandom sequential entry of subjects over time (accidental bias). Models for the assessment of these types of bias are presented in the preceding chapters.

Another objective of randomization is to permit a randomization-based analysis based on the exact or large sample probability distribution of a test statistic over the reference set of possible randomization permutations. Power is a property of hypothesis tests based on a population model, not a randomization model. However, we can see the impact on error rates when a population model is used to generate responses. Particularly interesting is the preservation of error rates when standard assumptions of tests under a population model are violated. We have seen, for instance, that under a linear time trend, the nominal type I error rate is preserved for Monte Carlo re-randomization tests.

It is important to remember, as discussed in Chapter 2, that randomization alone is not sufficient to provide an unbiased comparison of groups. Two other criteria

Randomization in Clinical Trials: Theory and Practice, Second Edition.
William F. Rosenberger and John M. Lachin.
© 2016 John Wiley & Sons, Inc. Published 2016 by John Wiley & Sons, Inc.

(Lachin, 2000) are required to ensure that a study result is unbiased. The first is that missing data from any randomized subjects, if any, do not bias the comparison of groups. This can be achieved by an intent-to-treat design in which all randomized subjects are followed so long as alive, able, and consenting. Second, the outcome assessments must be obtained in an equivalent and unbiased manner for all patients. The latter is obtained by double-masking or single-masking those who conduct the outcome assessments.

8.2 STRATIFICATION

The initial considerations in the design of a study stem from the study objectives. These should be stated in such a manner that specifies the target population, the number of treatment groups, the treatment regimens to be compared, the principal outcome measure to be used to compare the effects of treatment, the sample size per group, and the duration of follow-up. Given these, the first step in the implementation of randomization is to determine the number of strata, if any, to be employed in the randomization.

Methods of stratified analysis under a randomization model are discussed in Chapter 7. Standard methods may also be applied under a population model, such as the Mantel–Haenszel stratified analysis of multiple 2×2 tables (*cf.* Lachin, 2010). In Chapter 7, we showed that, while stratification of the randomization on a covariate will promote balance of that covariate within each randomized group, the statistical gains in efficiency resulting from the stratified randomization are indeed small compared to a stratified analysis using the same strata but without a stratified randomization. Thus, stratification must be justified on other grounds.

In most multicenter clinical trials, the greatest source of patient heterogeneity with respect to covariates for the particular disease under study is the clinical center in which the subjects are recruited. Further, there is often a wide range of numbers randomized within the various clinical centers. In addition, in some trials, poor-performing clinics may actually be dropped from the study, and it would be desirable that the elimination of the patients from one clinic does not affect the integrity of the randomization in the remaining clinics. For all these reasons, it is generally recommended that the randomization be stratified by clinical center or randomization site. (See Section 7.2.)

It may also be argued that one should also stratify on other major covariates. The motivation is to ensure that a major imbalance in such a covariate does not occur by chance among the treatment groups. However, as shown in Chapter 4, the probability of treatment imbalances is small, and any chance imbalances that do occur can effectively be adjusted for in the analysis. This assumes, however, that such covariate imbalances are due to chance among the cohort initially randomized and that there is complete follow-up of the randomized cohort. However, imbalances among groups selected from among those randomized, such as due to incomplete follow-up, cannot be adjusted for, with or without initial stratified randomization, without special unverifiable assumptions.

When stratification is deemed necessary for a large number of covariates, or if the total strata sizes are small, a covariate-adaptive randomization procedure can be employed. These procedures will be described in Chapter 9.

8.3 CHARACTERISTICS OF RANDOMIZATION PROCEDURES

In this section, we review important characteristics of the restricted randomization procedures described in Chapter 3. The basic issue is a trade-off between the desire to promote or guarantee balance in the numbers of treatment assignments versus the susceptibility to either selection bias or accidental covariate imbalances. In Chapter 6, we discussed preservation of error rates by the randomization test. This is another important criterion.

While Efron's maximum eigenvalue assessment of accidental bias is of academic interest, in practice, it is likely not a useful concept because it quantifies the magnitude of a severe bias, not its risk. Perhaps the only method that would be disfavored by Efron's accidental bias criterion is the truncated binomial design, without blocks or strata, since for this design, the maximum bias increases in n. In our view, susceptibility to accidental bias is not the most useful criterion for choosing one randomization procedure over another. Simply investigating the probability of a covariate imbalance leads to similar conclusions as an investigation of treatment imbalances.

In an unmasked study, or one with the potential for unmasking, the principal concern is the potential for introduction of selection bias. From our experience, it is human nature to try to arrange for a patient whom one feels is better suited to receive treatment $A(B)$ to be more likely to receive that treatment. This could be done, for example, by scheduling the randomization visit when one thinks it is more likely that the next assignment will be $A(B)$.

8.3.1 Consideration of selection bias

In a double-masked study with perfect allocation concealment, there is less susceptibility to selection bias. However, side effects or perceived differential response to treatment could lead to conclusions about the treatment assigned. In any event, most statisticians would agree that, all other factors being equal, the least predictable sequence is desirable. This leads to a metric of how close the conditional and unconditional treatment assignment probabilities are to each other, which in most cases is equivalent to the Blackwell–Hodges convergence strategy.

Given the stratum sample sizes, one should evaluate the probability of imbalances within strata and in aggregate with each procedure under consideration. In so doing, one should note that minor imbalances, such as 55 : 45 within strata or in aggregate, have little impact on the statistical properties of the study (although there may be ethical consequences). Although such imbalances may be of cosmetic concern, this alone should not require consideration of more aggressive restricted randomization procedures to ensure or promote balance. Other reasons, however, which may justify such an aggressive approach, are when the agent is a new drug therapy for which it is desired to ensure that approximately half the subjects are assigned to the active therapy to accrue the required patient-years of exposure for the assessment of side effects; or where the experimental therapy is very expensive or staff intensive and for budgetary reasons, balance is desired. In such studies with a small sample size in total, or within strata, complete randomization or urn designs may not be acceptable because there is a modest probability of such minor imbalances. Also, the probability of an extremely unusual sequence, such as *AAAA ... BBBB* or *ABABAB*, although small, may make these procedures undesirable for small sample sizes. Restricting the reference set appropriately, by choosing a restricted randomization procedure (such as those with imbalance intolerance), can therefore be important.

In an unmasked study, the potential for selection bias becomes a major concern. As shown in Chapter 5, the more predictable the sequence of assignments, or the higher the correlation among successive assignments, the greater the potential for clinic site staff to influence the composition of the treatment groups by scheduling patients for randomization visits so as to try to "beat" the randomization. The procedure most susceptible to such bias is the permuted block design with small block sizes. Under the Blackwell–Hodges model, using a convergent guessing strategy, random block sizes have little effect on the potential for such bias. Matts and Lachin (1988) explored an alternate model in which bias is introduced only when the block size is known and the investigator can predict with certainty or identify those assignments in the tail of a block where the probability of assignment to *A* or *B* is 1. In this case, the use of random block sizes will substantially reduce this potential for bias. In practice, neither model for selection bias is particularly accurate. In our experience, the greater the imbalance, the greater the temptation to attempt to beat the randomization. This leads to a model like that proposed by Stigler (1969). Regardless, it is clear that in an unmasked study, it is a temptation to try to beat the randomization, and during the course of a study, many clinic staff succumb.

Some additional advantage might be accrued by using a truncated binomial procedure rather than a random allocation rule to generate the assignments within each block. As shown in Chapter 5, sequences with lower potential for selection bias under the Blackwell–Hodges model have higher probability of occurrence using the truncated binomial than the random allocation rule. However, there are also the sequences with a longer tail of assignments with certainty; for example, the *AABB* sequence. Although not studied, one would conclude that the truncated binomial design would have greater potential for selection bias under the Matts–Lachin model of predictions with certainty when the block size is known. Thus, if the truncated binomial design is employed, use of random block sizes would be prudent.

With moderate strata or total sample sizes, however, the urn design is markedly preferred to the blocked designs due to the much lower susceptibility to selection bias. This approach has been used in a number of large-scale, unmasked multicenter trials with stratum sample sizes of 30 and higher including the Diabetes Control and Complications Trial (DCCT) (The Diabetes Control and Complications Trial Research Group, 1986) and The Diabetes Prevention Program (DPP) (Diabetes Prevention Program Research Group, 1999).

8.3.2   Implications for analysis

In Chapters 6 and 7, we describe methods of analysis based on the randomization distribution of a family of linear rank statistics with respect to the probabilities associated with each of the multitude of possible permutations. Such a randomization-based analysis has the advantage that it requires no assumptions other than the fact that a specific randomization procedure was employed and all observations were obtained in an unbiased manner. This is markedly different from the usual methods of analysis, such as a t-test for means or a chi-square test for proportions, which rely on the concept of sampling at random from a homogeneous population.

If it is planned that all analyses will be justified under population model assumptions, then the method of randomization is irrelevant to the choice of an analytic strategy or procedure. This is the approach most often taken in the analysis of a clinical trial and usually without controversy. However, in many instances, the variance of a test statistic can be markedly larger under population model assumptions than under randomization-based assumptions, and the resulting p-value is higher.

Of course, the two approaches are assessing different questions. The randomization analysis addresses the probability that a difference at least as large as that observed among the n subjects randomized into the trial could have occurred by chance under the null hypothesis. The population model analysis addresses the probability that such a difference could have been observed in samples of n_A and n_B subjects drawn at random from their respective populations. Thus, the randomization analysis allows a conclusion about the effects of treatment among the n patients studied, whereas the population model analysis allows a conclusion, through confidence interval estimation, about the effects of the treatment in a general hypothetical population from which the n patients were drawn.

As we stated in Chapter 6, in our opinion, both approaches have value. We think it prudent to specify in the protocol that a randomization analysis would be used to conduct a test of the treatment group effect on the primary outcome and that a population model analysis would be used to estimate the effect within the general population.

8.4   SELECTING A RANDOMIZATION PROCEDURE

The restricted randomization procedures we have described have only been compared with respect to an individual criterion separately. However, the properties for determining a "good" randomization procedure involve multiple criteria and competing objectives. For example, perfect balance can be attained by the deterministic sequence *ABABAB* ..., but this is the most predictable sequence under the Blackwell–Hodges model. Alternatively, complete randomization is completely unpredictable, but it has the worst balancing properties. Many authors have suggested using graphics to explore these trade-offs (e.g., Zhao *et al.*, 2011; Atkinson, 2014; Baldi Antognini, and Zagoraiou, 2014). In this section, we present *trade-off plots* that graph competing criteria simultaneously. Such graphics can be used to choose the best parameters of a design for a given sample size, such as the bias p of the biased coin design (Atkinson, 2014) or compare many different procedures simultaneously.

An assumption of the trade-off plot is that, given the same scaling, a one unit change in the x direction is equivalent in importance to a one unit change in the y direction. Such an assumption is very unrealistic, but trade-off plots still have value in comparing many procedures simultaneously. One can change the scaling if one criterion is considered more important than the other. In this section, we assume equal importance of criteria.

Scaling the criteria to $(0, 1)$ is rather simple for most criteria. For the imbalance metric, we can divide $\text{Var}(D_n)$ by n. This would scale the metric to $(0, 1)$, with complete randomization taking the worst value 1. This scaled metric has a special interpretation. As in Burman (1996), we can think of this metric as the expected effective number of patients on whom information is lost due to the imbalance of the design under a normal model with homoscedastic variances. For the expected selection bias factor in (5.2), we divide by $n/2$.

8.4.1   Choosing parameter values

We can use trade-off plots to find appropriate parameter values for a randomization procedure. Figures 8.1 and 8.2 compare the trade-off between predictability, as measured by the expected selection bias factor divided by $n/2$, and balance, as measured by $\text{Var}(D_n)/n$ for different parameters of the same procedure when $n = 50$. The criteria are simulated 100,000 times as in Tables 3.4 and 5.1.

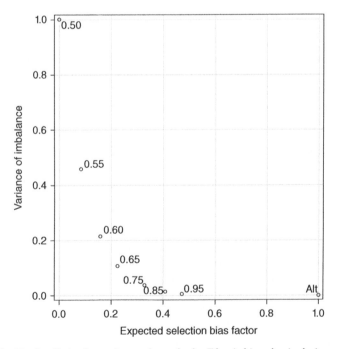

Fig. 8.1   *Trade-off plot for various values of p for Efron's biased coin design, n = 50.*

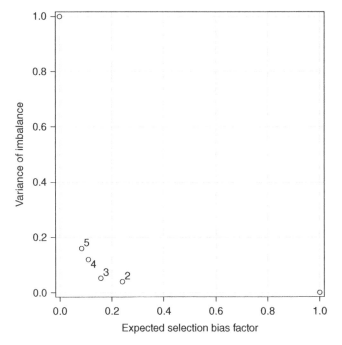

Fig. 8.2 *Trade-off plot for various values of b for the big stick design, n = 50.*

Figure 8.1 varies the value of p for Efron's biased coin design. It appears that Efron's conjecture of $p = 2/3$ as being a suitable value for the biased coin design is well-founded. Certainly, somewhere around 0.64 is the point on the plot closest to the origin. Baldi Antognini *et al.* (2015) derive the value of p that optimizes weighted compound criteria exactly for balance versus selection bias and balance versus accidental bias. When the two criteria are weighted equally, the trade-off plots allow visualization of the formal optimality problem. They present a trade-off plot in three dimensions, with the weights as the third dimension.

For the big stick design in Figure 8.2, an imbalance intolerance bound of $b = 3$ or 4 seems to be the best choice, with $b = 3$ favoring lower variability and $b = 4$ favoring lower predictability.

8.4.2 Comparing procedures

Trade-off plots are also useful in comparing restricted randomization procedures, and they can facilitate the appropriate selection of procedures for different criteria. For example, Figures 8.3 and 8.4 plot the variance of the imbalance against the expected bias selection factor for 12 procedures. At one extreme is complete randomization ("CR"), where $\mathrm{Var}(D_n)/n = 1$ and the expected selection bias factor is 0; at the other extreme is a deterministic alternating sequence $ABAB \ldots$ ("Alt"), which has $\mathrm{Var}(D_n)/n = 0$ and expected selection bias factor 1.

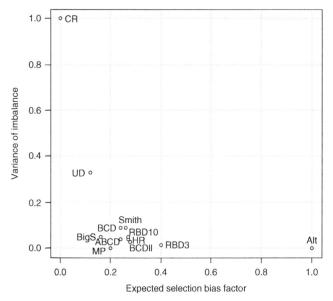

Fig. 8.3 *Trade-off plot for various restricted randomization procedures, n = 50, comparing imbalance and predictability measures.*

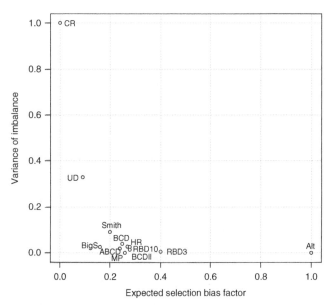

Fig. 8.4 *Trade-off plot for various restricted randomization procedures, n = 100, comparing imbalance and predictability measures.*

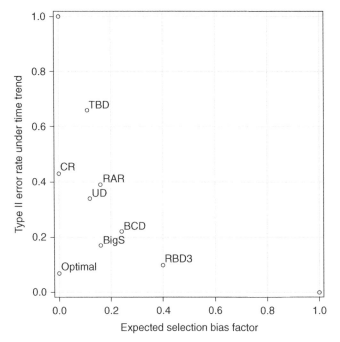

Fig. 8.5 *Trade-off plot for various restricted randomization procedures, n = 50, comparing predictability and type II error rate under a linear time trend.*

In between are the restricted randomization procedures described in Chapter 3: Wei's UD(0, 1) ("UD"), Smith's design ("Smith") with $\rho = 5$, Efron's biased coin design ("BCD", $p = 2/3$), the big stick design ("BigS") and biased coin design with imbalance intolerance ("BCDII") (both with $b = 3$), the accelerated biased coin design ("ABCD") with $a = 10$, the maximal procedure ("MP") with $b = 3$, Hadamard randomization ("HR") using the matrix in (3.13), and the random block design with $B_{\max} = 3, 10$ ("RBD3", "RBD10"), filled using the random allocation rule. For both sample sizes, the big stick design is the only procedure appearing in the lower left quadrant and, therefore, is the "best" design with respect to the trade-off between imbalance and predictability, if one weights the measures equally.

In Figure 8.5, we compare the procedures with respect to the trade-off between the expected selection bias factor and the type II error rate of the randomization test under time trends ($n = 50$) for the procedures in Table 6.4. The procedures are depicted using the same acronyms as in Figures 8.3 and 8.4. The point "optimal" has expected selection bias factor 0 and type II error rate 0.07, which represents the nominal 93% power achieved by the t-test under complete randomization and homogeneous $N(0, 1)$ responses. Again, using these criteria, the big stick design appears to be the "best" procedure.

8.4.3 Conclusions

Trade-off plots are a useful way to compare randomization procedures when there are competing objectives in the trial. While each trial has its own unique considerations, and some criteria may be more important than others, there are some interesting observations that can be made based on the limited investigations we have conducted. First, while the permuted block design or random block design is used in the vast majority of clinical trials, there does not appear to be any advantage to using a blocking strategy. In every comparison we have conducted, other restricted randomization procedures can be found that outperform the blocked designs. Secondly, the big stick design seems to perform the best of all randomization procedures with respect to the competing objectives we have discussed in this book. While we have looked only at small sample sizes, $n = 50, 100$, most large multicenter clinical trials would likely be stratified by center, and so these results may be relevant even for large trials.

8.5 GENERATION OF SEQUENCES

Standard computer packages such as SAS and R have built-in random number generators that can be used to prepare the randomization sequence for a clinical trial. Using the notation of Chapter 3, a loop over the n patients will generate the assignments plus increment the function $N_A(j), j = 1, \dots, n$. For most restricted randomization rules described, the probability of assignment to A, given by $p_{Aj} = E(T_j | \mathcal{F}_{j-1})$, is simply a function of $N_A(j - 1)$. For each patient, a uniform random number $U_j \in [0, 1]$ is generated, and the following rule is applied:

$$\text{If } U_j \leq p_{Aj}, \text{ assign treatment } A$$
$$U_j > p_{Aj}, \text{ assign treatment } B.$$

It should be noted that randomization sequences are only as good as the random number generator. One should test any random number generator used with appropriate goodness-of-fit statistics. The RANUNI function in SAS is highly regarded as a reliable random number generator.

For the case of three groups, say A, B, and C, this approach generalizes by determining the desired probability of each treatment allocation. Let the probability that patient j will be assigned to A be denoted as p_{Aj} and the probability that patient j will be assigned to B be denoted as p_{Bj}. Based on the random uniform number U_j, the jth subject is then assigned as follows:

$$A \quad \text{if} \quad U_j \leq p_{Aj};$$
$$B \quad \text{if} \quad p_{Aj} < U_j \leq (p_{Aj} + p_{Bj});$$
$$C \quad \text{if} \quad U_j > (p_{Aj} + p_{Bj}).$$

This approach immediately generalizes to the case of any number of treatments.

A random number should be specified as the seed to initialize the sequence. The random number seed and the edition/version of the random number generator should be documented so that the sequence can be regenerated and replicated if need be.

One frequent question from experimenters is whether one need to actually draw a random permutation as the basis for the randomization, as opposed to simply writing down an "attractive" sequence of assignments or using a systematic sequence such as *ABAB* …. While the answer should be clear from the preceding chapters, it is helpful to explore the historical views on this. It has long been recognized that such "attractive" sequences are likely to be somewhat systematic, if not formally so, and that systematic sequences are susceptible to systematic biases. That is, the characteristics of the sample units differ in a systematic way, which corresponds to the systematic differences in assignments, resulting in substantial bias. More importantly, due to their nonrandomness, such designs often fail to provide an accurate measure of residual error or an accurate reflection of the unexplained random variation. In R.A. Fisher's (1935) treatise *The Design of Experiments* that many consider the basis for formal randomization in experimentation, he first points out (p. 63) that

> … the results of using arrangements which differ from the random arrangement … are thus in one way or another undesirable since they will tend to underestimate or overestimate the true residual error.

Then in a discussion of systematic versus random Latin squares, he states (p. 77) that

> The failure of systematic arrangements came not from recognizing that the function of the experiment was not only to make an unbiased comparison, but to supply at the same time a valid test of its significance. This is vitiated equally whether the components affecting the comparisons are larger or smaller than those on which the estimate of error is based.

Despite the desire to achieve true "randomness," it is common practice that randomization sequences are examined and perhaps rejected and replaced if the sequence is considered undesirable. If the only consideration were the cosmetic properties of a sequence, then such rejection and re-randomization would be warranted. However, this practice violates the assumptions required for a randomization-based analysis, which is based on the probabilistic structure over the complete reference set of possible permutations. The use of an appropriate restricted randomization procedure should eliminate the possibility of generating a wildly discrepant sequence, and procedures that have an imbalance intolerance level, or the random block design, eliminate the possibility of these sequences.

8.6 IMPLEMENTATION

In most instances, the random assignments are generated prior to the start of recruitment into a study and then a system specified for the implementation of the randomization as the subjects are recruited. This system of randomization will also provide for the single- or double-masking of the assignments.

8.6.1 Packaging and labeling

For a double-masked pharmaceutical trial, it is necessary that a supply of placebo material be provided that is indistinguishable from the active agent with respect to appearance, consistency, touch, weight, taste, smell, and so on. Patients or their companions are often tempted to open a capsule or crush a pill to see if they can detect the presence/absence of the active agent. During clinic visits, patient may compare the weight of their medication supplies. The same also applies to clinic staff. Thus, the placebo and active material should be as equivalent as possible in all respects other than the active agent.

In a clinical trial of a surgical or intervention procedure, masking is implemented by use of a sham procedure for those randomized to the control group. In the extreme case of a major surgical procedure, this would include a trip to the operating room with anesthetization, incision, and wound closure. Clearly, in most cases, this would be considered unethical. In other less extreme cases, such as a minor procedure under local anesthesia, or an infusion, or use of radiation, and so on, a sham may be acceptable.

Randomization in pharmaceutical trials is implemented differently from intervention or surgical trials because of differences in the nature of the therapies. The simplest method for the implementation of a pharmaceutical trial is to employ randomization by lots or supplies. In this case, the physician or pharmacist has two lots of material labeled 1 or 2. If a patient is assigned to A (or B), then a supply of medication is drawn from lot 1 (or 2) and provided to the patient, or the physician/practitioner, for administration. This approach is fine for an unmasked study. However, this is a poor approach for a masked study because if any one patient's treatment assignment is unmasked, due to an adverse event, or overdose, or whatever, then the entire study is unmasked. A variation would be to use a blocked-lot procedure, such as randomization by lots within clinical centers. In this case, unmasking of an individual study subject would unmask the assignments within that block, but not necessarily unmask the entire study, provided that the lot 1 and 2 contents were varied within sites.

The most secure way of implementing a double-masked randomization for a pharmaceutical trial is to provide a unique supply of medication, prepackaged and labeled, for each individual subject. This could be bottles of medication (or other containers) with preassigned patient numbers or supply numbers. One approach is to assign each patient a unique randomization number at the time of randomization and to have a prepared supply of medication for that randomization ready for dispensation or administration at the time of randomization. For example, in a multicenter trial, patient numbers might be assigned of the form $ccxxx$ where cc is the clinical site number ($01, 02, \ldots$) and xxx is the patient number within clinical site ($001, 002, 003, \ldots$). When the patient is randomized, and assigned a unique randomization number, a unique supply of study material is assigned to that patient. In this way, emergency or inadvertent unmasking of a single patient has no impact on the integrity of the masking of other patients.

This system, however, requires that each study subject be identified by two study numbers. Prior to randomization, subjects undergo a period of screening to assess eligibility to enter the study and perhaps even a trial period of treatment with a placebo to assess compliance with the medication regimen. In some cases, patient must also be withdrawn from pre-existing medications or be stabilized using a specified regimen. Thus, a study number is assigned at the initiation of screening and another number assigned at the time of randomization.

A variation on this approach is to simply have a set of study supplies packaged and labeled according to one numbering scheme and a system to assign a supply number to a patient at the time of randomization. In this case, a patient might be assigned a study screening number at the time of the initial screening visit and then a study supply number at the time of randomization. However, the number of A and B supplies required is unknown in advance, and so there must be an oversupply of drug available.

For an unmasked trial pharmaceutical trial, a pharmacist can administer the medication using a supply of the active material, those assigned to control receiving nothing. For a surgical or intervention trial, the clinic staff need only know whether a subject is assigned to the experimental or control arms. In such cases, subjects are assigned a study number at the time of initial screening and then later assigned to treatment A or B.

8.6.2 The actual randomization

The treatment assignment can be conveyed to the clinical sites in a variety of ways. The oldest type of system, and in some respects the least favorable, is a system of sealed envelopes. Envelopes labeled by study patient number are distributed to the sites, and as each successive patient is randomized, the next envelope is opened and the enclosed study supply or intervention group is assigned to that patient. This system should never be used because it allows all the envelopes to be opened in advance, thus potentially unmasking the sites to the sequence of assignments and opening the study to extreme selection bias.

Rather a system that guarantees that future assignments remain unknown should be implemented. One approach is to employ central randomization where the clinical site contacts a central center to either verify the randomization in a double-masked study or provide the treatment assignment in an unmasked study. In either case, it is advisable that there be central verification of the randomization process. It is recommended that, prior to actual randomization, the central office verify that the subject meets the entrance eligibility criteria (with no exclusions), has consented to randomization and full study participation, and is ready to administer the treatment immediately. Even in the case where a prepackaged supply of medication or study material is ready for assignment in the clinic, it is advisable that central verification be employed. In studies without these checks, patients have been randomized as those who did not meet all eligibility criteria, who did not consent, or who did not ever receive any study medication. This is inexcusable. In a simple pharmaceutical trial, the recommended procedure is that the clinical site call a central office, verify that the patient is eligible and consenting, then have that patient's supply of medication ready for administration. The physician then meets with the patient, opens the bottle, removes a pill, hands the patient a glass of water, and asks the patient to swallow the pill. If the patient balks, then the patient is not randomized into the study. In this case, the bottle of medication can be returned to the pharmacy for destruction, but the same patient number (supply) is assigned to the next willing patient.

Such central verification can be provided by a telephone call to central staff or by a central computer facility. In the latter case, the program asks the site to answer a series of questions using the key pad and then verifies or provides the random assignment. Another approach is to provide an interactive website for this purpose.

Finally, there must also be some system for postrandomization emergency unmasking. In most pharmaceutical trials, separate patient supplies of medication are prepared by a central distribution pharmacy based on the randomization sequence generated by the study statistician. The pharmacy should have operating procedures to ensure that each treatment (e.g., active versus placebo) is processed in a manner to ensure that the randomization is executed as specified. This could include random selection of the packaged material for testing by a simple technique such as a litmus test. The central pharmacy should also provide a 24-hour answering service to answer any questions about the contents of each medication supply, and all supplies should be labeled with information about this service. In some studies, the pharmacy may be instructed to notify a study monitor, such as the study medical monitor, of the unmasking.

8.7 SPECIAL SITUATIONS

The earlier discussion applies to a simple two (or more) group study design. There are studies, however, that present additional considerations.

In some studies, two (or more) active agents are to be used, such as where a new agent is compared to an active control, often manufactured or supplied by a different company. In this case, it is not possible to provide an identical formulation for the two agents such that the study material is identical with respect to appearance, taste, and so on, for each group. Then in order to maintain masking, a *double placebo* approach is necessary. Supplies of each active agent and a placebo for each agent are prepared, and each subject is asked to take two pills, one from bottle A and another from bottle B. If the patient is assigned to receive active treatment A (or B), then the supply of medication labeled A (or B) contains the active agent and the supply for the other agent labeled B (or A) contains the placebo. The patient takes two pills at a time, one containing active agent, the other a placebo.

This technique might be necessary in a two-group, positive controlled effectiveness trial or an equivalence trial. A generalization is a 2×2 factorial design where patients are assigned to receive either control, A alone, B alone, or A and B in combination. In this case, the A and B bottles contain placebo A and placebo B, active A and placebo B, placebo A and active B, and active A and active B, respectively.

In some cases, the A and B supplies also differ in form. For example, the A therapy may be administered orally as a capsule and the B agent by infusion. Regardless, masking can be preserved by administration of the matching placebo of the other agent to patients in either group.

In some cases, one may have a design with an untreated control group and two active therapy groups where each therapy has a different formulation, requiring a double placebo approach to maintain complete double-masking. However, if one of the agents is administered orally, and the other by infusion, then a complete double placebo implementation may not be ethical. For example, suppose that m subjects are assigned to each of treatments A (active oral), B (active infusion), and C (neither). Patients and local Institutional Review Boards may object to the administration of a sham (placebo) infusion to the $2m$ subjects assigned to treatments A or C. However, incomplete double-masking may be obtained by randomly assigning half the control patients to receive the A placebo and half to receive the sham B infusion. In this manner, m subjects receive A and $m/2$ the A placebo, and m subjects receive B and $m/2$ the B sham. This would be implemented by a two-stage randomization. First, patients are randomly assigned to receive treatments A, B, or C. Then those assigned to receive C are randomly assigned to also receive either the A placebo or the sham B infusion. Note that there are only three groups for purposes of statistical analysis: the subjects who receive the A placebo and the sham B infusion combine to form the C control group.

Finally, randomization procedures may be implemented so as to "share" controls in multiple parallel protocols. For example, suppose that a study is launched using A versus A placebo. Later another study is initiated in an identical population using identical procedures to compare B versus B placebo. For the period while the two studies overlap, an incompletely double-masked, three-arm randomization could be employed as earlier, where half the subjects assigned to the "control" group are then randomly assigned to receive the A placebo and half the B placebo. However, all subjects assigned to group C during the period of overlap would be included in the analysis of the A versus control in the A study and also in the analysis of the B versus control in the B study. Thus, the subjects in group C are contained in the control group for both the A study and the B study.

Consider the case where study A alone is recruiting over some period, followed by the simultaneous recruitment to the A and B studies during a second period, followed by the close of recruitment to study A and continued recruitment to study B during a third period. During the first and third periods, a simple two-group randomization is employed. In the middle period, a three-group randomization is employed to A, B, or C. In order to maintain total double-masking, half those assigned to receive C during this middle period could then be assigned to receive either the A placebo or the B placebo via a supplemental randomization. This three-group randomization should be implemented in such a way that the parallel two-group randomizations are not affected.

For a permuted block design with m assignments to each treatment per block, where m may vary among blocks, then during the first and third periods, balanced blocks of length $2m$ are employed, while during the second period, blocks of length $3m$ are employed. Let n_A, n_B, and n_C represent the number of patients assigned to A, B, and C, respectively. For a biased coin design, the two-group randomization in the A study would assign to group A with probability p_A when $n_C > n_A$ and likewise in the B study to group B with probability p_B when $n_C > n_B$. For example,

Table 8.1   *Biased coin allocation ratios to A, B, and C such that the probability of assignment to A is $p_A = 2/3$, and to B is $p_B = 2/3$, when there is an excess number of prior allocations to C.*

Imbalance	$A : B : C$
$n_A > n_B > n_C$	$1 : 1 : 2$
$n_A > n_C > n_B$	$1 : 4 : 2$
$n_B > n_A > n_C$	$1 : 1 : 2$
$n_B > n_C > n_A$	$4 : 1 : 2$
$n_C > n_A > n_B$	$2 : 2 : 1$
$n_C > n_B > n_A$	$2 : 2 : 1$

consider the case where $p_A = p_B = 2/3$ so that the biased coin allocations are in either a $2:1$ or a $1:2$ ratio, depending on whether the excess allocations in the past are to control or active treatment, respectively. Then during the second period when the allocations for the A and B studies are performed simultaneously, the possible settings and the corresponding allocation ratios are shown in Table 8.1. These allocations will preserve the desired $2:1$ ratio for assignments to A versus C and for B versus C, but there is no control over A versus B imbalance during the period of joint allocations.

A similar strategy can be employed for an urn $UD(\alpha, 1)$ design. Initially, for the first period where only the A study is recruiting, the urn contains α balls of type A and of type C. At the end of this period, at the start of the second period, let n_{1A} and n_{1C} refer to the numbers of allocations to A and to C, respectively. Thus, the urn contains $\alpha + n_{1C}$ balls of type A and $\alpha + n_{1A}$ balls of type C. Then to initialize the randomization to also include study B, $\alpha + n_{1A}$ balls of type B are also added to the urn. This equals the number of C balls in the urn so that B and C are allocated with equal probability on the next draw. After each draw, a ball is added to the urn for each of the two types other than that drawn. At the end of the second period, at the conclusion of recruitment to study A, the urn contains $\alpha + n_{1A} + n_{2A} + n_{2C}$ balls of type B, and $\alpha + n_{1A} + n_{2A} + n_{2B}$ balls of type C, where n_{2A}, n_{2B}, and n_{2C} are the numbers of allocations made to each treatment during the second period. At this point, all A balls are removed from the urn, as well as the excess B and C balls from the initial period and due to the A allocations during the second period, leaving $\alpha + n_{2C}$ balls of type B and $\alpha + n_{2B}$ balls of type C. These are the numbers of balls that would have been in the urn had one started with randomization only to B and C, which produced n_{2B} allocations to B and n_{2C} to C. For a $UD(\alpha, \beta)$ design, the aforementioned "n"values would be multiplied by β.

This design would also tend to balance the A to B allocations during the second phase, in addition to balancing the A to C and B to C allocations. This approach would also be used to continue allocations in a multigroup study after one of the arms has been discontinued, such as due to adverse events, as illustrated in the DPP example as follows.

8.8 SOME EXAMPLES

8.8.1 The optic neuritis treatment trial

Fifteen clinical centers enrolled 457 patients using a permuted block design with a separate sequence for each clinical center. Patients were randomly assigned to one of three treatment regimens: intravenous methylprednisolone, oral prednisone, or oral placebo. Whereas the patients in the oral-prednisone and placebo groups were not informed of their treatment assignments, those in the intravenous-methylprednisolone group were aware of their assignments. The primary outcome was the development of multiple sclerosis, there being a significant reduction in risk among those assigned to steroids. (See Beck *et al.*, 1993.)

8.8.2 Vesnarinone in congestive heart failure

In a preliminary study of the drug vesnarinone in the treatment of congestive heart failure, two clinics each recruited 40 subjects who were assigned to receive double-masked vesnarinone versus placebo. A permuted block design with block size of 2 was used. The study showed a reduction in mortality among these 80 subjects. (See Feldman *et al.*, 1993.)

8.8.3 The Diabetes control and complications trial

The DCCT enrolled 1441 subjects with type 1 (juvenile) diabetes within 29 clinical centers who were randomly assigned to receive either intensive versus conventional therapy for the control of blood glucose levels. The study showed that the intensive group, which maintained lower levels of blood glucose, had significantly reduced risk of microvascular complications of diabetes (Diabetes Control and Complications Trial Research Group, 1986, 1993). The DCCT enrollment was conducted in two stages. In the initial feasibility stage, a total of 278 patients were recruited in 23 clinics during 1983–1998. The randomization was stratified by adults versus adolescents within each clinical center, 46 strata total. Due to the small sample size per clinic, and the requirement that each clinic enroll at least four adolescents, an initial permuted block of four subjects were assigned within each stratum, followed by a UD(0, 1) randomization. In the second stage of recruitment from 1984–1989, six new clinics were added and the randomization was stratified by a primary prevention cohort (no pre-existing retinopathy) versus a secondary intervention cohort (some pre-existing mild retinopathy, among other differences) within clinical center, 58 strata total. To initialize the urns for each strata within the original 23 clinics, the 278 assignments were poststratified by primary/secondary cohort and clinical center. Then the appropriate number of balls of each type were placed in the urn for each stratum. For example, if there were four intensive and three conventional patients within the primary cohort of a given clinic from the feasibility phase, then three intensive and four conventional balls were placed in the urn for that stratum. The sequences allowing for 50 subjects within each of the 58 strata were then generated with a specified seed. For the six new clinics, the sequences started with zero balls in each of their 12 strata. Each sequence was inspected to ensure that there were no long runs of assignments to either treatment. In one stratum, a run of 9 *A*s was followed by a run of 7 *B*s. One element from each run was randomly selected to be changed to the other treatment. Of the 1441 patients randomized into the study, in the primary prevention cohort, 378 were assigned to receive conventional therapy, 348 intensive therapy; and in the secondary intervention cohort, 352 were assigned to receive conventional therapy, 363 intensive therapy.

8.8.4   Captopril in diabetic nephropathy

Thirty clinical centers recruited 409 subjects with pre-existing diabetic nephropathy. Using a UD(0, 1) procedure stratified by clinical center, a total of 207 were assigned to receive double-masked captopril and 202 to receive placebo. The risk of further progression of diabetic nephropathy was reduced by 48 percent with captopril. (See Lewis *et al.*, 1993.)

8.8.5   The diabetes prevention program

In the DPP, a total of 27 clinical centers recruited 3234 adult subjects with impaired glucose tolerance who were followed to observe the incidence of type 2 diabetes. Using a UD(0, 1) procedure stratified by clinical center, subjects were assigned to receive either lifestyle intervention aimed at weight loss through diet and exercise, or conventional lifestyle management plus the drug troglitazone, or conventional management plus the drug metformin, or conventional management plus placebo. The lifestyle treatment was unmasked. In order to maintain masking among the three medication treatment groups, a double placebo technique was employed where each subject took two pills daily containing either one of the active agents or placebo. The troglitazone arm was terminated due to adverse effects after 585 patients had been randomized to receive troglitazone. These patients were unmasked and their treatment terminated. The remaining subjects were then told only to take their assigned pills from the metformin bottle, half containing active agent, half placebo. At that point, the composition of the urn was modified to shift from a four-arm randomization to a three-arm randomization. All the troglitazone balls were removed from the urn, as well as the 585 balls of each other type added due to these prior troglitazone allocations. At the end of the study, 1079 subjects had been assigned to lifestyle therapy, 1073 to metformin, and 1082 to metformin-placebo. The study showed that lifestyle intervention achieved approximately a 58 percent reduction in the risk of developing diabetes versus placebo, whereas metformin yields a 31 percent risk reduction. (See Diabetes Prevention Program Research Group, 1999.)

8.8.6   Scleral buckling versus primary vitrectomy in retinal detachment (The SPR Study)

A randomized clinical trial in patients with rhegmatogenous retinal detachments was conducted to compare two surgical techniques: scleral buckling versus primary vitrectomy (the "SPR Study"; see Heimann *et al.*, 2001). Two separately conducted parallel trials, each of size $n = 200$, were conducted in phakic and aphakic patients, due to differences in their postoperative course. A stratified randomization was employed using the random block design with block sizes 2 and 4, with 49 strata representing the 49 different surgeons participating. Strata with less than 10 patients per parallel trial were collapsed for analysis purposes; a population-based analysis was conducted. Due to the surgical nature of the procedure, masking was impossible, but outcome measures were assessed by a masked endpoint committee.

8.9 PROBLEMS

8.1 Prepare a trade-off plot similar to that in Figure 8.1 for the following procedures and draw conclusions about appropriate values of the parameter:

a. the random block design for $B_{max} = 2, 3, \ldots , 10$.
b. Smith's design for $\rho = 0, 1, \ldots , 5$.

8.2 Prepare a trade-off plot similar to that in Figure 8.3 for $n = 200$. Do results change?

8.3 For each of the examples in Section 8.8, discuss the properties of the randomization procedure employed with respect to the potential for selection bias, accidental bias, or other biases, and the implications for a randomization versus population model analysis.

8.4 Would complete randomization, with or without stratification as appropriate, be an acceptable approach in the DCCT? Justify your answer.

8.5 In each of the following cases, describe the randomization procedure you would employ and justify your answer. Generate the procedure and describe the resulting sequence.

(i) An investigator is planning a phase II trial with only 100 patients recruited in 10 clinical centers, with a range of 8–12 expected per center. The study will be double-masked and employ an active treatment versus control.

(ii) Consider an equivalent study but where the treatments by nature must be administered in an unmasked manner.

(iii) Now consider a Phase III study where 1000 patients are to be recruited in 20 clinical centers, with a range of 30–70 within each center. The trial will be double-masked and employ an active treatment versus control.

(iv) Consider an equivalent study but where the treatments by nature must be administered in an unmasked manner.

(v) Now consider a Phase III study where 1000 patients are to be recruited in 50 clinical centers, with a range of 10–30 within each center. The trial will be double-masked and employ an active treatment versus control.

(vi) Consider an equivalent study but where the treatments by nature must be administered in an unmasked manner.

8.6 In the vesnarinone study, since the block size used was 2, subjects were randomly assigned within pairs. Would a randomization using a larger block size be acceptable? Justify your answer.

8.7 In the captopril study in diabetic nephropathy, approximately 25 percent of subjects entered the study with significant loss of renal function as represented by a serum creatinine exceeding 1.5 mg/dL. A stratified analysis was planned among those with such high values and among those with lower values. Should the randomization have also been stratified by high versus low initial creatinine values? Justify your answer.

8.8 As is now mandated for all major studies launched by the National Institutes of Health, one of the objectives of the Diabetes Prevention Program was to assess the effects of treatment among ethnic subgroups, both genders, and the elderly. The study recruitment targets included the recruitment of 50 percent ethnic minorities, including African Americans, Native Americans, Asian-Americans, and Hispanics, both genders, and 20 percent of subjects of at least 60 years of age. Should the randomization have also been stratified by any of these factors? Justify your answer.

8.10 REFERENCES

ATKINSON, A. C. (2014). Selecting a biased-coin design. *Statistical Science* **29** 144–163.

BALDI ANTOGNINI, A., ROSENBERGER, W. F., WANG, Y., and ZAGORAIOU, M. (2015). Exact optimum coin bias in Efron's randomization procedure. *Statistics in Medicine*, in press.

BALDI ANTOGNINI, A. and ZAGORAIOU, M. (2014). Balance and randomness in sequential clinical trials: the dominant biased coin design. *Pharmaceutical Statistics* **13** 119–127.

BECK, R. W., CLEARY, P. A., TROBE, J. D., KAUFMAN, D. I., KUPERSMITH, M. J., PATY, D. W., and BROWN, C. H. (1993). The effect of corticosteroids for acute optic neuritis on the subsequent development of multiple sclerosis. The Optic Neuritis Study Group. *New England Journal of Medicine* **329** 1764–1769.

BURMAN, C. F. (1996). *On Sequential Treatment Allocation in Clinical Trials*. Sweden: University of Göteborg (doctoral dissertation).

Diabetes Control and Complications Trial Research Group. (1986). The Diabetes Control and Complications Trial (DCCT): design and methodological considerations for the feasibility phase. *Diabetes* **35** 530–545.

Diabetes Control and Complications Trial Research Group. (1993). The effect of intensive treatment of diabetes on the development and progression of long-term complications in insulin-dependent diabetes mellitus. *New England Journal of Medicine* **329** 977–986.

Diabetes Prevention Program Research Group. (1999). Design and methods for a clinical trial in the prevention of type 2 diabetes. *Diabetes Care* **22** 623–634.

FELDMAN, A. M., BRISTOW, M. R., PARMLEY, W. W., CARSON, P. E., PEPINE, C. J., GILBERT, E. M., STROBECK, J. E., HENDRIX, G. H., POWERS, E. R., BAIN, R. P., WHITE, B. G., and The Vesnarinone Study Group. (1993). Effects of Vesnarinone on Morbidity and Mortality in Patients with Heart Failure. *New England Journal of Medicine* **329** 149–155.

FISHER, R. A. (1935). *The Design of Experiments*. Oliver and Boyd Edinburgh.

HEIMANN, H., HELLMICH, M., BORNFELD, N., BARTZ-SCHMIDT, K.-U., HILGERS, R.-D., and FOERSTER, M. H. (2001). Scleral buckling versus primary vitrectomy in rhegmatogenous retinal detachment (SPR Study): design issues and implications. *Graefe's Archives of Clinical and Experimental Ophthalmology* **239** 567–574.

LACHIN, J. M. (2000). Statistical considerations in the intent-to-treat principle. *Controlled Clinical Trials* **21** 167–189.

LACHIN, J. M. (2010). *Biostatistical Methods: The Assessment of Relative Risks.* John Wiley & Sons, Inc., New York.

LEWIS, E. J., HUNSICKER, L. G., BAIN, R. P., and ROHDE, R. D. (1993). The effect of angiotensin-converting-enzyme inhibition in diabetic nephropathy. *New England Journal of Medicine* **329** 1456–1462.

MATTS, J. P. and LACHIN, J. M. (1988). Properties of permuted block randomization in clinical trials. *Controlled Clinical Trials* **9** 327–344.

STIGLER, S. M. (1969). The use of random allocation for the control of selection bias. *Biometrika* **56** 553–560.

ZHAO, W., WENG, Y., WU, Q., and PALESCH, Y. (2011). Quantitative comparison of randomization designs in sequential clinical trials based on treatment balance and allocation randomness. *Pharmaceutical Statistics* **11** 39–48.

9

Covariate-Adaptive Randomization

In the previous chapters, we have assumed that a set of s strata is defined on the basis of one or more covariates and a separate randomization is performed within each stratum. An entirely different approach would be to determine the treatment assignment of a new subject so as to minimize the covariate imbalances within treatment groups. This approach has been called by various names in the literature: *adaptive stratification*, *dynamic allocation*, or *minimization*. When randomization is involved in the procedure, we prefer the term *covariate-adaptive randomization*.

9.1 EARLY WORK

Some of the early work in this area appeared in the 1970s in papers by some of the great researchers in clinical trials: Zelen, Pocock, Simon, Taves, and Wei. It appears that, with the exception of Wei (1978), which references the Pocock and Simon (1975) paper, these authors were unaware of the other authors' work, despite the similarity in intentions.[1]

Except for Zelen's approach, all of these procedures attempt to induce balance between treatment groups with respect to covariate margins and are therefore called *marginal approaches*. One can immediately see the benefit of such an approach if there are a large number of stratification variables. For example, if there are 8 binary covariates, there will be 256 strata, while the number of covariate margins is only 16. The marginal approach is not concerned with treatment balance in each stratum, but only across each covariate separately.

[1] All three papers were published in 1974 or 1975, so it is difficult to establish priority, but the submission dates put Taves at February 1973, Pocock and Simon at November 1973, and Zelen at November 1973. Both Zelen and Pocock were at SUNY Buffalo at the time.

Randomization in Clinical Trials: Theory and Practice, Second Edition.
William F. Rosenberger and John M. Lachin.
© 2016 John Wiley & Sons, Inc. Published 2016 by John Wiley & Sons, Inc.

9.1.1 Zelen's rule

Zelen's (1974) rule uses a preassigned randomization sequence (which could be generated by complete randomization or some restricted randomization design) ignoring strata. Let $N_{ik}(n)$ be the number of patients in stratum $i = 1, \ldots, S$ on treatment $k = 1, 2$ $(1 = A, 2 = B)$. When patient $n + 1$ in stratum i is ready to be randomized, one computes $D_i(n) = N_{i1}(n) - N_{i2}(n)$. For an integer c, if $|D_i(n)| < c$, then the patient is randomized according to schedule; otherwise, the patient receives the opposite treatment. Zelen proposes $c = 2, 3$, or 4. He also proposes randomizing the value of c for each new patient.

9.1.2 The Pocock–Simon procedure

Pocock and Simon (1975) proposed a randomized version of Zelen's rule. Let $N_{ijk}(n), i = 1, \ldots, I, j = 1, \ldots, n_i, k = 1, 2$ $(1 = A, 2 = B)$, be the number of patients in category j of covariate i on treatment k after n patients have been randomized. (In our previous notation, $\prod_{i=1}^{I} n_i = S$ is the total number of strata in the trial.) Suppose the $(n + 1)$th patient to be randomized is a member of categories $r_1, \ldots, r_I$ of covariates $1, \ldots, I$. Again, we define D to be a difference metric; in this case, let $D_i(n) = N_{ir_i1}(n) - N_{ir_i2}(n)$. We then take a sum over weighted strata defined by $D(n) = \sum_{i=1}^{I} w_i D_i(n)$, where w_i are weights chosen depending on which covariates are deemed of greater importance. If $D(n)$ is less than 0, then the weighted difference measure indicates that B has been favored thus far for that set, $r_1, \ldots, r_I$, of strata and the patient $n + 1$ should be assigned with higher probability to treatment A, and vice versa, if $D(n)$ is greater than 0. Pocock and Simon suggest biasing a coin with

$$p = \frac{c^* + 1}{3} \tag{9.1}$$

and implementing the following rule: if $D(n) < 0$, assign the next patient to treatment A with probability p; if $D(n) > 0$, assign the next patient to treatment A with probability $1 - p$; and if $D(n) = 0$, assign the next patient to treatment A with probability $1/2$, where $c^* \in [1/2, 1]$.

Note that if $c^* = 1$, we have a rule very similar to Efron's biased coin design in Section 3.5. If $c^* = 2$, we have the deterministic *minimization method* proposed by Taves (1974). Many other rules could be considered, all derivatives of Zelen's rule and Taves's minimization method with a biased coin twist to give added randomization; Efron (1980) describes one such rule and applies it to a clinical trial in ovarian cancer.

Pocock and Simon generalize their covariate-adaptive randomization procedure for more than two treatments by considering a general metric $D_i^k(n), k = 1, \ldots, K$, which could be the standard deviation of the $N_{ir_ik}(n)$'s, and a weighted sum $D^k(n) = \sum_{i=1}^{I} w_i D_i^k(n)$. The D^k's are then ordered from smallest to largest, and a corresponding set of probabilities $p_1 \geq p_2 \geq \ldots \geq p_K$ is determined such that $\sum_{k=1}^{K} p_k = 1$. The value $p_k, k = 1, \ldots, K$, is then the probability that patient $n + 1$ with strata $r_1, \ldots, r_I$

will be assigned to treatment k. Pocock and Simon suggest the following functional form:

$$p_k = c^* - \frac{2(Kc^* - 1)}{K(K + 1)}k, \; k = 1, \dots, K. \tag{9.2}$$

Note that (9.2) reduces to (9.1) for $K = 2$.

9.1.3   Example: Adjuvant chemotherapy for locally invasive bladder cancer

In a multicenter clinical trial of patients with locally invasive bladder cancer (Stadler *et al.*, 1993), who have undergone a radical cystectomy with lymph node dissection, and whose tumors demonstrate p53 abnormalities, 190 patients were to be randomized to either adjuvant chemotherapy (95 patients) or routine follow-up (95 patients) after surgery. The primary outcome was time to recurrence. Because of the large number of stratification variables, it was decided that the Pocock–Simon procedure would be used (see Section 9.1.2). Stratification variables were age (dichotomized at 65 years), stage (dichotomous), grade (dichotomous), and p21 status (dichotomous). The value of p in (9.1) was set to 0.75 ($c^* = 1.25$). The trial was stopped for futility after 114 patients were randomized. All baseline covariates included in the randomization procedure were balanced, with a maximum imbalance between treatment groups of 6 in the over 65 years old category.

9.1.4   Wei's marginal urn design

Wei (1978) described the use of an urn model for covariate-adaptive randomization. When the number of covariates is such that the resulting number of strata is large and the stratum sizes are small, using a separate urn in each stratum can result in imbalances in treatment assignments within strata. Let n_i be the number of categories for the ith of I covariates considered jointly, such that there are $s = \prod_{i=1}^{I} n_i$ unique strata. Instead of using s urns, one for each unique stratum, Wei proposed using an urn for each category of each covariate, for a total of $\sum_{i=1}^{I} n_i$ urns. For a given new subject with covariate values $r_1, \dots, r_I$, the treatment group imbalances within each of the corresponding urns are examined. The one with the greatest imbalance is used to generate the treatment assignment. A ball from that urn is chosen and then replaced. Then β balls representing the opposite treatment are added to the urns corresponding to that patient's covariate values. Wei called this a *marginal urn*.

9.1.5   Is marginal balance sufficient?

Marginal balance may seem to be inadequate compared with balance in each stratum, but this is not necessarily so. Wei (1978), in a remarkably cogent short proof, shows that, if there is no interaction among the covariates nor between the treatment effect and covariates, marginal balance is sufficient. Consider a linear model where Y is the response of a patient to treatment $k = 1, \dots, K$ that has covariate category r_i for factor $i, i = 1, \dots, I$. We can write the standard ANOVA model as

$$Y = \mu + \sum_{i=1}^{I} \beta_{ij_i} + t_k + \epsilon, \tag{9.3}$$

where t_k and β_{ij_i} are treatment and covariate category effects, μ is a constant, and ϵ is a $N(0, \sigma^2)$ error term, with the usual constraint that $\sum_{j=1}^{n_i} \beta_{ij} = 0, i = 1, \dots, I$.

Suppose we wish to estimate a treatment contrast $\sum_{k=1}^{K} c_k t_k$ for a K-dimensional contrast vector c. If we have equal numbers of patients in each stratum, then the best linear unbiased estimator is $\sum_{k=1}^{K} c_k \bar{Y}_k$, where $\bar{Y}_k$ is the observed mean of Y in treatment group k. If we do not have equal numbers in each stratum, then the mean squared error of $\sum_{k=1}^{K} c_k \bar{Y}_k$ can be computed as

$$\text{MSE}\left(\sum_{k=1}^{K} c_k \bar{Y}_k\right) = \sum_{k=1}^{K} \frac{c_k^2 \sigma^2}{n_k} + \left\{\sum_{k=1}^{K} c_k \left(\sum_{i=1}^{I} \sum_{j_i=1}^{n_i} \frac{n_{ij_i k}}{n_k} \beta_{ij_i}\right)\right\}^2, \tag{9.4}$$

where n_k is the number of patients assigned to treatment k and $n_{ij_i k}$ is the number of patients assigned to k in category j_i of covariate i. We can see that if $n_{ij_i k}/n_k$ is constant for $k = 1, \dots, K$, which means equal allocation proportions within covariate margins across treatment arms, then the bias term is 0. Hence, marginal balance is sufficient to ensure that the treatment contrast estimator is unbiased.

If we have interactions, either among covariates or among covariates and treatment, then one needs to have balance within all covariate-by-treatment strata. In the next section, we describe some covariate-adaptive procedures that attempt to do this.

9.1.6 Is randomization necessary?

Taves's (1974) minimization procedure is not randomized, and there was never any intention in the original paper of using any form of randomization. Rather minimization was proposed as an alternative to randomization. Hidden in the discussion of that paper is a very cogent heuristic argument about why minimization can never result in a distribution of unknown covariates that is worse than from a randomized clinical trial.

Consider an important but neglected or unknown variate: either it does or does not have some degree of correlation with the variates used for minimizing. If it does have some correlation, either positive or negative, it will tend to be distributed like the variates with which it is correlated. If the variate is not correlated it will be placed in the two groups independently of anything done to the minimized variates, i.e., placed randomly. To have the full range of a random distribution it must be possible to have extremely skewed combinations. Since minimizing works against these skewed combinations theoretically–and practically, ... , the odds for skewed combinations with minimization will never be greater than with randomization.

Aickin (2001) has argued that randomization is not needed in covariate-adaptive procedures because the covariates themselves are random, leading to randomness in the treatment assignments. In addition, for large multicenter clinical trials with centralized randomization and minimization, he believes the likelihood of correctly guessing the treatment assignment of the next patient would be minimal, thus eliminating concerns about selection bias. Thirty years after proposing minimization, Taves (2004) still supports the view that randomization is not necessary, writing:

> I hope the day is not too far distant when we look back on the current belief that randomization is essential to good clinical trial design and realize that it was ... "credulous idolatry".

Berger (2010) and Senn believe that selection bias and predictability are the major concerns with deterministic minimization. Barbáchano, Coad, and Robinson (2008) show that, under deterministic minimization, scant information about the last treatment assignment can lead to more than 60 percent correct guesses, and this increases to 80 percent if the last two assignments are known. Berger (2010) states:

> ... The idea behind minimization is brilliant, but its failure is the flip side of the same coin, and cannot be separated from its benefit. The claim to fame is that subjects are allocated not randomly but rather deterministically, so as to minimize an imbalance function (hence the name). It is this, the heart and soul of the method (and not some tangential aspect), that leads to its downfall .... If the odds of each treatment come close enough to 50% so as to truly take a bite out of selection bias, then the baby is lost with the bath water, and minimization no longer does what it purports to do. There is simply no way around this.

Senn (2013) writes:

> ... The advantage of minimisation compared with randomisation is a potential gain in efficiency. However, it is also known that the gain is very small and dependent on a number of assumptions. The disadvantages include an increase in predictability with its attendant disadvantages (loss of blinding vulnerability to manipulation) and increased uncertainty about precision itself. My judgement is that this particular game is not worth the candle, and I strongly prefer randomisation to minimisation.

9.2 MORE RECENT COVARIATE-ADAPTIVE RANDOMIZATION PROCEDURES

In this section, we review more recent covariate-adaptive randomization procedures. We first describe methods to balance within strata, marginally, and overall for a set of discrete strata. We then describe methods that treat covariates as continuous.

9.2.1 Balancing within strata

One of the criticisms of the methods described thus far based on marginal balance across covariates is that, within individual strata, imbalances can result. While this has been shown to not be problematic if there are no interactions, some authors have developed techniques to balance within strata as well as marginally. Signorini *et al.* (1993) developed *dynamic balanced randomization* based on Soares and Wu's big stick design. First, a hierarchical ordering of the stratification structure is defined. For example, a clinical trial with two covariates, gender and clinic, say, the hierarchy may be gender within clinic ($i = 1$), then clinic ($i = 2$), then the overall trial ($i = 3$). Each of these I hierarchical stratification levels has an associated imbalance intolerance, $b_i, i = 1, \ldots, I$. Randomization is conducted with probability $1/2$ unless one of the imbalance intolerance bounds is reached, where b_1 is examined first, then b_2, until b_I. If an imbalance intolerance bound is attained, then the treatment assigned less often is allocated with probability 1.

In our example with two covariates, and two clinics, if the next patient to be randomized is a female in clinic 2, then the difference between treatment assignments among females in clinic 2 is compared to b_1. If it exceeds the threshold, then the treatment assigned less often so far is assigned with probability 1. Otherwise, one examines the imbalance in clinic 2 and compares it to the threshold. If it exceeds the threshold, then the treatment assigned less often so far is assigned with probability 1. Otherwise, one looks at the treatment imbalance overall in the trial and compares it to the threshold b_3. If it exceeds the threshold, then the treatment assigned less often so far is assigned with probability 1. Otherwise, the patient is randomized with probability $1/2$.

An alternative method is to create a compound distance metric that is a weighted sum of the imbalance within strata, within margins of covariates, and overall (Hu and Hu, 2012). Suppose the $(n + 1)$th patient enters the trial with covariate values $(r_1, \ldots, r_I)$ of covariates $(1, \ldots, I)$. For each treatment k that could be assigned, say $k = A, B$, we compute three sets of difference metrics. Let $D_{\text{stratum},k}(n)$ be the squared difference of patient numbers on treatment A versus treatment B among the first n patients whose covariate profile is $(r_1, \ldots, r_I)$. Let $D_{ik}(n)$ be the squared difference of patient numbers among the first n patients whose ith covariate value is r_i. Let $D_k(n)$ be the squared difference of patient numbers in the trial after n patients. Define a set of $I + 2$ nonnegative weights, $(w_{\text{margin},i}, i = 1, \ldots, I, w_{\text{trial}}, w_{\text{stratum}})$ that sum to 1. We compute an imbalance score, $G_k, k = A, B$ as follows:

$$G_k(n) = \sum_{i=1}^{I} w_{\text{margin},i} D_{ik}(n) + w_{\text{trial}} D_k(n) + w_{\text{stratum}} D_{\text{stratum},k}(n). \tag{9.5}$$

Efron's biased coin design is then used to assign the next treatment, depending on the value of $G_A(n) - G_B(n)$. The $n + 1$th patient is then assigned to treatment A with probability p if $G_A(n) - G_B(n) < 0$, with probability $1 - p$ if $G_A(n) - G_B(n) > 0$, and with probability $1/2$ if $G_A(n) - G_B(n) = 0$, where $p \in (1/2, 1)$.

9.2.2 Balancing with respect to continuous covariates

All of the methods discussed thus far have discretized covariates to create strata. Hu *et al.* (2015) review nine papers that attempt to balance continuous covariates across treatments. Some of the procedures are designed to induce comparability with respect to means, medians, variances, or ranks. Hu, *et al.* mention that the ultimate goal would be to balance the actual distribution of the covariates across treatments. One method of doing this is to define a metric based on a comparison of the areas under the empirical distribution function among treatments. Another possible metric compares the kernel density estimators across treatments. All of these procedures rely on the same technique of defining a distance metric and then performing a biased coin randomization to favor the treatment that will promote greater balance. While feasible in practice, it is not clear that such procedures actually perform better than covariate-adaptive randomization on discretized covariates.

An alternative approach when there are continuous (and/or categorical) covariates is to find the treatment that maximizes the efficiency of estimating the treatment effect in a covariate-adjusted regression model, as described in the next section.

9.3 OPTIMAL DESIGN BASED ON A LINEAR MODEL

The covariate-adaptive randomization procedures in the preceding sections are arbitrary in the sense that they are developed intuitively rather than based on some optimal criterion. While one can simulate these procedures for different parameter values and find appropriate designs to fit certain criteria, none of the designs has been shown to be optimal. Instead of concerns about balance of treatment assignments across strata, one can take an entirely different approach and find an allocation rule that minimizes the variance of the estimated treatment effect in the presence of covariates. Such a rule would necessarily require the specification of a model linking the covariates and the treatment effect.

Begg and Iglewicz (1980) and Atkinson (1982) select a standard linear regression model. Here we follow Atkinson's development. We begin with the classical regression model, given by

$$E(Y_i) = x_i'\beta, i = 1, \ldots, n,$$

where the Y_i's are independent responses with $\text{Var}(Y) = \sigma^2 I$ and x_i includes a treatment indicator and selected covariates of interest. Then $\text{Var}(\hat{\beta}) = \sigma^2 (X'X)^{-1}$, where $X'X$ is the $p \times p$ dispersion matrix from n observations.

For the construction of an optimal design, we wish to find the n points of experimentation at which some function is optimized (in our case we will be finding the optimal sequence of n treatment assignments). The dispersion matrix evaluated at these n points is given by $M(\xi_n) = X'X/n$, where ξ_n is the n-point design. It is convenient, instead of thinking of n points, to formulate the problem in terms of a measure ξ (which is, in this case, a frequency distribution) over a design region Ξ.

Since an important goal of clinical trials is to estimate a treatment effect, possibly adjusting for important covariates, Atkinson formulates the optimal design problem as a design that minimizes, in some sense, the variance of $A'\beta$, where A is an $s \times p$

matrix of contrasts, $s < p$. One possible criterion is Sibson's (1974) D_A-optimality that maximizes

$$|A'M^{-1}(\xi)A|^{-1}. \tag{9.6}$$

Other criteria could also be applied. Atkinson compares the D_A criterion to standard *D-optimality*, which maximizes the log determinant of M. Ball, Smith, and Verdinelli (1993) investigate the *Bayesian D-optimality criterion*, where a Bayesian prior distribution is assumed for β, and the procedure maximizes the expectation (with respect to the prior distribution) of the log determinant of M.

For any multivariable optimization problem, we compute the directional derivative of the criterion. In the case of the D_A criterion in (9.6), we can derive the directional (Frechet) derivative as

$$d_A(x, \xi) = x'M^{-1}(\xi)A(A'M^{-1}(\xi)A)^{-1}A'M^{-1}(\xi)x.$$

By the classical equivalence theorem of Kiefer and Wolfowitz (1960), the optimal design ξ^* that maximizes the criterion (9.6) then satisfies the following equations:

$$\sup_{x \in \Xi} d_A(x, \xi) \leq s, \forall \xi \in \Xi \tag{9.7}$$

and

$$\sup_{x \in \Xi} d_A(x, \xi^*) = s \tag{9.8}$$

(Kiefer and Wolfowitz (1960); see Atkinson and Donev (1992) for further details.)

In the model with covariates, we have

$$E(Y) = x_1\beta_1 + x_2'\beta_2,$$

where x_1 is the treatment indicator vector and x_2 is a vector of important covariates. In this case, if we are interested in estimating the treatment effect in the presence of covariates, $A' = [A_1' : 0]$ with A_1 identifying the treatment differences. This formulation can be simplified with two treatments, but the optimal design that satisfies (9.7) and (9.8) must be determined numerically.

Such a design is optimal for estimating linear contrasts of β, but the solution will provide only an allocation ratio to each of K treatments, without incorporating the sequential nature of a clinical trial. Assume that n patients have already been allocated, and the resulting n-point design is given by ξ_n. Atkinson proposes a sequential design that allocates the $(n + 1)$th patient to the treatment $k = 1, \ldots, K$ for which $d_A(k, \xi_n)$ is a maximum. However, this design is deterministic.

In order to randomize the allocation, Atkinson suggests biasing a coin with probabilities

$$p_k = \frac{d_A(k, \xi_n)}{\sum_{k=1}^{K} d_A(k, \xi_n)} \tag{9.9}$$

and allocating to treatment k with the corresponding probability. With two treatments, $k = 1, 2$ $(1 = A, 2 = B)$, we have $s = 1, A' = [-1, 1]$, and the probability of assigning treatment A is given by

$$p = \frac{d_A(1, \xi_n)}{d_A(1, \xi_n) + d_A(2, \xi_n)}. \tag{9.10}$$

With no covariates in the model, the model becomes $E(Y) = \beta_k, k = 1, 2$, and the equations in (9.7) and (9.8) can be solved analytically. We can write (9.10) as

$$p = \frac{\{N_B(n)\}^2}{\{N_A(n)\}^2 + \{N_B(n)\}^2}, \tag{9.11}$$

where $N_A(n)$ and $N_B(n)$ are the numbers of patients assigned to treatments A and B, respectively, through n patients (Problem 4.4). Note that this is the design in equation (3.25) with $\rho = 2$. In a similar vein, Ball, Smith, and Verdinelli (1993) and Atkinson (1998) investigate Bayesian optimality criteria.

While Atkinson's (1982) original approach relies on a homoscedastic linear model. In principle, one could extend these results to any generalized linear models, such as logistic regression or parametric survival models, but the optimal solution will depend on the unknown parameters of the model. There are three options for dealing with this problem. One can substitute a best guess of the parameter values to obtain a *locally optimal design*. Secondly, one can take a Bayesian approach by putting a prior distribution on the unknown parameters and obtain a *Bayesian optimal design*. Finally, one could substitute sequentially computed estimates of those parameters based on the response data accrued thus far. Such models are called *covariate-adjusted response-adaptive* (CARA) designs and will be discussed in the next chapter. (The topic is dealt with more extensively in Chapter 9 of Hu and Rosenberger (2006)).

9.4   THE TRADE-OFF AMONG BALANCE, EFFICIENCY, AND ETHICS

There is no doubt that minimization and covariate-adaptive randomization tend to balance a set of known covariates marginally across treatments and can thus improve the cosmetic appearance of comparisons of baseline characteristics (often found in Table 1 of clinical trials reports). This has been shown in countless simulation papers that will not be described here. But is balance on covariates really essential in the valid comparison of treatments in a clinical trial, or is its primary importance only cosmetic? This has been the subject of much controversy. Certainly if the patient cohorts on different treatment arms have vastly different characteristics, the trial can be less convincing. While covariate-adjusted treatment effects can be estimated using post-hoc regression analysis, many would argue that heterogeneity should be eliminated entirely, through either stratification or covariate-adaptive randomization. The same arguments used for and against stratification, given in Chapter 7, apply to covariate-adaptive randomization.

Rosenberger and Sverdlov (2008) describe the problem in terms of a trade-off among competing objectives of the clinical trials, shown in Figure 9.1. As Atkinson (1982) showed, in the event that the treatment effect arises from a homogeneous linear regression model, the optimal design for efficient estimation of the treatment contrast is one that balances marginally over covariates. However, this model is not applicable for nonlinear models with heterogeneous variances, such as logistic regression or survival models (e.g., Begg and Kalish, 1984; Kalish and Harrington, 1988). In these cases, balancing marginally over covariates can actually reduce the efficiency of estimation. These issues have been hotly debated in the literature; see Atkinson (1999) and the ensuing discussion.

The third component in Figure 9.1 that has yet to be mentioned is ethical considerations. While balance may be more efficient in limited cases, balanced allocation can lead to more patients being assigned to the inferior treatment. This may be problematic, particularly in trials where the outcome is grave. We will discuss this issue further in the next chapter, along with randomization methods that may mitigate the problem. Rosenberger and Sverdlov (2008) give several examples where balance in treatment numbers and balance marginally over covariates are less efficient and less ethically appealing than imbalances. They agree with Senn that " ... cosmetic balance, while psychologically reassuring, should not be the goal if power or efficiency is lost in the process of forcing balance."

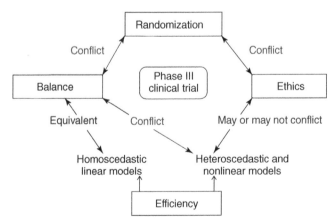

Fig. 9.1 *Multiple objectives of a phase III clinical trial. Source: Rosenberger and Sverdlov, 2008, p. 411. Reproduced with permission of Institute of Mathematical Statistics.*

9.5   INFERENCE FOR COVARIATE-ADAPTIVE RANDOMIZATION

In this section, we describe techniques to analyze data following covariate-adaptive randomization. We begin with model-based inference and then address randomization-based inference.

9.5.1   Model-based inference

Shao, Yu, and Zhong (2010) derive a sufficient condition for a hypothesis test to be valid under covariate-adaptive randomization when a model is specified among the responses and a set of covariates. Let $(\mathcal{T}, \mathcal{Y}, \mathcal{X})$ be the set of treatment assignments, responses, and covariates, respectively, for n patients. A hypothesis test S is *valid* for a significance level α if, under H_0,

$$\lim_{n\to\infty} P_{\mathcal{Y},\mathcal{T}}(|S| > c_\alpha|\mathcal{X}) \le \alpha, \tag{9.12}$$

with equality holding for at least some cases, and $P_{\mathcal{Y},\mathcal{T}}(\cdot|\mathcal{X})$ denotes a probability calculated with respect to the conditional distribution of $(\mathcal{Y},\mathcal{T})$, given $\mathcal{X}$. If theallo-cation is fixed, so that $\mathcal{T}$ is nonrandom, and a regression model between $\mathcal{Y}$ and $\mathcal{X}$ is correctly specified, then

$$\lim_{n\to\infty} P_{\mathcal{Y}}(|S| > c_\alpha|\mathcal{X}) \le \alpha, \tag{9.13}$$

so that S is valid under any deterministic allocation. For example, if $E(Y_i|X_i)$ is linear in X_i, then the one-way analysis of covariance test is valid under any fixed allocation, provided $E(Y_i|X_i)$ is correctly specified. If we now incorporate randomization, an application of the dominated convergence theorem gives the equality

$$\lim_{n\to\infty} P_{\mathcal{Y}}(|S| > c_\alpha|\mathcal{X}, \mathcal{T}) = E_{\mathcal{T}} \left\{ \lim_{n\to\infty} P_{\mathcal{Y}}(|S| > c_\alpha|\mathcal{X}, \mathcal{T})|\mathcal{X} \right\}.$$

Hence, a sufficient condition for a test S satisfying (9.13) to be valid is

$$P_{\mathcal{Y}}(|S| > c_\alpha|\mathcal{X}, \mathcal{T}) = P_{\mathcal{Y}}(|S| > c_\alpha|\mathcal{X}), \tag{9.14}$$

or that $\mathcal{T}$ and $\mathcal{Y}$ are conditionally independent, given $\mathcal{X}$. This condition holds under complete randomization and under any restricted randomization procedure. However, it does not necessarily hold under covariate-adaptiverandomization.

However, under covariate-adaptive randomization, if the set of covariates used in the design, $\mathcal{Z}$ is a function of $\mathcal{X}$, the test is still valid. This is because

$$P_{\mathcal{Y}}(|S| > c_\alpha|\mathcal{X}, \mathcal{T}) = P_{\mathcal{Y}}(|S| > c_\alpha|\mathcal{X}, \mathcal{Z}, \mathcal{T})$$
$$= P_{\mathcal{Y}}(|S| > c_\alpha|\mathcal{X}, \mathcal{Z})$$
$$= P_{\mathcal{Y}}(|S| > c_\alpha|\mathcal{X}),$$

as $\mathcal{T}$ and $\mathcal{Y}$ are conditionally independent, given $(\mathcal{X}, \mathcal{Z})$. Hence, for valid inference following covariate-adaptive randomization, the covariates $\mathcal{Z}$ must be a function of the model covariates $\mathcal{X}$ and the model must be correctly specified. Forsythe (1987) among others have recommended that all covariates used in the covariate-adaptive randomization should also be used in the analysis, and hypothesis tests will be valid if this is followed, provided the model for analysis is correctly specified.

This result is sufficient, but not necessary. A number of authors have shown simulation evidence that hypothesis tests can be conservative if $\mathcal{Z}$ is left out of the analysis (e.g., Birkett, 1985; Forsythe, 1987; Barbáchano and Coad, 2013). Shao, Yu, and Zhong (2010) give a theoretical justification of this phenomenon. Consider the following linear model linking $\mathcal{Y}, \mathcal{T}$, and $\mathcal{Z}$:

$$Y_{ij} = \mu_j + \beta Z_i + \epsilon_{ij}, \ i = 1, \ldots, n, j = A, B,$$

where the Z_i's are i.i.d. with finite second moment, μ_j and β are unknown parameters, and ϵ_{ij} are i.i.d random errors with mean zero, variance σ^2, independent of the Z_i's. They show that, unless $\beta = 0$ in the model, the t-test under covariate-adaptive randomization is conservative, due to inducedcovariance between the sample means $\bar{Y}_A$ and $\bar{Y}_B$; that is, there exists a constant α_0 such that, under H_0,

$$\lim_{n \to \infty} P_{\mathcal{Y},\mathcal{T}}(|S| > c_\alpha) \leq \alpha_0 < \alpha.$$

Senn (2004) gives a more philosophical view of why the treatment effect should be adjusted for the covariates used in the design:

> If an investigator uses [covariate-adaptive randomization], she or he is honour bound, in my opinion, as a very minimum, to adjust for the factors used to balance, since the fact that they are being used to balance is an implicit declaration that they have prognostic value. In the case of a linear model the standard error quoted will generally be too small if this is not done . . . .

In conclusion, valid inference under a population model can only be attained under covariate-adaptive randomization if (1) a covariate-adjusted analysis is used, adjusting for the variables used for the randomization and (2) the model is correctly specified. Usual population-based inference methods valid under fixed, complete, or restricted randomization will be conservative otherwise.

9.5.2 Randomization-based inference

Randomization-based inference is more appropriate than population-based inference (see Chapter 6), and the technique of re-randomization for the Pocock–Simon method was described by Simon (1979, p. 508):

> It is possible, though cumbersome, to perform the appropriate randomization test generated by a nondeterministic adaptive stratification design. One assumes that the patient responses, covariate values, and sequence of patient arrivals are all

fixed. One then simulates on a computer the assignment of treatments to patients using the [Pocock–Simon procedure] and the treatment assignment probabilities actually employed. Replication of the simulation generates the approximate null distribution of the test statistic adopted, and the significance level. One need not make the questionable assumption that the sequence of patient arrivals is random.

Computation of a re-randomization test, while fixing the covariate values and responses, is no longer "cumbersome" due to today's computational resources, and the establishment of a randomization p-value follows the procedure described in Section 6.9. However, the randomization distribution is completely determined by the random sequences generated by the biased coin; with deterministic minimization, there is no randomization distribution (cf., Kuznetsova, 2010). Hasegawa and Tango (2009) show that a randomization test of the difference of means is equivalent to the analysis of covariance when there is marginal balance in covariates. In that case, the randomization test is more powerful and less conservative than the t-test.

Proschan, Brittain, and Kammerman (2011) and Simon and Simon (2011) show that the re-randomization test preserves the type I error rate in the event of time trends. While, in principle, unequal allocation can be targeted using covariate-adaptive randomization, randomization-based hypothesis tests can be subject to a serious loss of power compared to tests under normal theory in a population model (cf., Proschan, Brittain, and Kammerman, 2011; Kuznetsova and Tymofyeyev, 2013).

9.6 CONCLUSIONS

Covariate-adaptive randomization can be an effective technique to force balance marginally on a set of known covariates. Whether marginal balance is essential in most clinical trials has not been resolved. In many cases, balance may be cosmetic and may not improve efficiency of estimation, particularly in the context of heteroscedastic, nonlinear models. Another approach is based on the theory of optimal designs; while more complicated to implement, it attempts to maximize the efficiency of estimation of the treatment effect. In any event, we favor those procedures that are randomized, to mitigate selection bias and to provide a basis for inference. These include the Pocock–Simon procedure and its variants, as well as Atkinson's biased coin procedure in (9.9). If such a procedure is used, it is important to adjust for the covariates in any population-based analysis. However, the advantages of randomization-based inference in such a context should be clear: the type I error rate is preserved in the presence of time trends, and valid inference does not depend on a correctly specified model.

There is an important logistical distinction between restricted and stratified randomization and covariate-adaptive randomization. For restricted and stratified randomization, the entire randomization sequence can be generated in advance, which aids in prepackaging treatments. For covariate-adaptive randomization, the randomization of each patient is not available until the patient's covariate profile is known.

9.7 PROBLEMS

9.1 Consider a two-arm clinical trial in which there are three baseline covariates over which balance is sought: (Z_1, Z_2, Z_3) representing gender, age, and cholesterol level. Assume that $Z_{ij}, i = 1, \ldots, n, j = 1, 2, 3$ are independently distributed as Bernoulli$(1/2)$, discrete uniform $[18, 75]$, and normal$(\mu = 200, \sigma = 30)$, respectively. Furthermore, assume that Z_2 and Z_3 are discretized into 4 and 3 levels, respectively:

$$\tilde{Z}_2 = \begin{cases} 1 \text{ (young)}, & \text{if } Z_2 \in [18, 30]; \\ 2 \text{ (older young)}, & \text{if } Z_2 \in [31, 45]; \\ 3 \text{ (middle age)}, & \text{if } Z_2 \in [46, 59]; \\ 4 \text{ (elderly)}, & \text{if } Z_2 \in [60, 75], \end{cases}$$

and

$$\tilde{Z}_3 = \begin{cases} 1 \text{ (normal level)}, & \text{if } Z_3 < 200; \\ 2 \text{ (borderline risk)}, & \text{if } Z_3 \in [200, 240); \\ 3 \text{ (high risk)}, & \text{if } Z_3 \geq 240. \end{cases}$$

Conduct a simulation study to describe the operating characteristics of the Pocock–Simon procedure in terms of balance (overall and within-covariate levels) and randomness for various levels of p.

9.2 Consider the model in (9.3).

a. Show that, if there are equal numbers in each stratum, the best linear unbiased estimator (BLUE) of $\sum_{k=1}^{K} c_k t_k$, for a K-dimensional contrast vector c, is $\sum_{k=1}^{K} c_k \bar{Y}_k$, where $\bar{Y}_k$ is the observed mean of Y in treatment group k.
b. Derive equation (9.4).

9.3 Consider the covariate-adaptive randomization procedure based on the imbalance score, described in (9.5). Suppose there are $J = 2$ covariates: clinic (1, 2, 3) and gender (male, female), and 100 patients have already been randomized according to the following table:

Gender	Treatment	Clinic 1	Clinic 2	Clinic 3
Male	A	11	7	6
	B	11	7	7
Female	A	9	9	8
	B	11	8	7

Assume that weights are $(w_{trial}, w_{margin,1}, w_{margin,2}, w_{stratum}) = (0.2, 0.2, 0.2, 0.4)$, and $p = 0.85$. Compute the probability that patient 101, a female in clinic 1, will be randomized to treatment A. (Hu *et al.*, 2015.)

9.4 [*Maximum entropy constrained balance*, Klotz (1978)] Consider a trial with $K \geq 2$ treatments. Let $B_i \geq 0$ be some measure of "overall covariate imbalance" from assigning the new patient to treatment i and $\mathbf{B} = (B_1, \ldots, B_K)$. Let $\mathbf{P} = (P_1, \ldots, P_K)$ be a vector of randomization probabilities to treatments $1, \ldots, K$. The expected covariate imbalance is

$$E(\mathbf{B}) = \sum_{i=1}^{K} B_i P_i.$$

One measure of randomness of treatment assignments is *entropy*, defined as

$$H(\mathbf{P}) = -\sum_{i=1}^{K} P_i \log P_i.$$

Consider the following optimization problem:

$$\begin{aligned} \max_{\mathbf{P}} \quad & H(\mathbf{P}) \\ \text{s.t.} \quad & E(\mathbf{B}) \leq \eta B_{(1)} + (1 - \eta)\bar{B} \\ \text{and} \quad & 0 \leq P_i \leq 1, \ \sum_{i=1}^{K} P_i = 1. \end{aligned} \tag{9.15}$$

In (9.15), $B_{(1)} = \min_{1 \leq i \leq K} B_i$, $\bar{B} = K^{-1} \sum_{i=1}^{K} B_i$, and $0 \leq \eta \leq 1$ is a user-specified constant that determines trade-off between complete randomization ($\eta = 0$) and strict balance ($\eta = 1$).

Using Lagrange multiplier optimization, show that the optimal randomization probabilities are as follows:

$$P_i = \frac{e^{-\lambda B_i}}{\sum_{k=1}^{K} e^{-\lambda B_k}}, \quad i = 1, \ldots, K,$$

where $\lambda = \lambda(\eta)$ is chosen such that $\sum_{i=1}^{K} B_i P_i = \eta B_{(1)} + (1 - \eta)\bar{B}$.

9.5 [*Minimum quadratic distance constrained balance*, Titterington (1983)] Let $\mathbf{P} = (P_1, \ldots, P_K)$ and $\mathbf{P}_B = (K^{-1}, \ldots, K^{-1})$. The quadratic distance between $\mathbf{P}$ and $\mathbf{P}_B$ is

$$\delta(\mathbf{P}, \mathbf{P}_B) = \sum_{i=1}^{K} (P_i - K^{-1})^2.$$

Consider the following optimization problem:

$$\begin{aligned} \min_{\mathbf{P}} \quad & \delta(\mathbf{P}, \mathbf{P}_B) \\ \text{s.t.} \quad & E(\mathbf{B}) \leq \eta B_{(1)} + (1 - \eta)\bar{B} \\ \text{and} \quad & 0 \leq P_i \leq 1, \ \sum_{i=1}^{K} P_i = 1. \end{aligned} \tag{9.16}$$

a. Using Lagrange multiplier optimization, show that the optimal randomization probabilities are as follows:

$$P_i = \frac{1}{K} + \eta \frac{(B_{(1)} - \bar{B})}{\sum_{k=1}^{K} (B_k - \bar{B})^2}, i = 1, \dots, K. \tag{9.17}$$

b. Show that when $K = 2$, (9.17) simplifies to

$$P_1 = (1 + \eta)/2, \quad P_2 = (1 - \eta)/2 \quad \text{if } B_1 < B_2;$$
$$P_1 = (1 - \eta)/2, \quad P_2 = (1 + \eta)/2 \quad \text{if } B_1 > B_2,$$

which is equivalent to Efron's biased coin design with coin bias $(1 + \eta)/2$.

9.6 Derive equation (9.11).

9.7 Consider a classical linear regression model with constant variance:

$$Y_n = Z_n \beta + \alpha t_n + \epsilon, \quad \epsilon \sim N(0, \sigma^2 I),$$

where Y_n is a vector of responses from n patients in the trial, α is the treatment effect, β is a $p \times 1$ vector of covariate effects, t_n is a $n \times 1$ vector of treatment assignments (1 for A; -1 for B), and Z_n is a $n \times p$ matrix of covariate values (including **1**, the intercept). Let $\theta = (\beta, \alpha)'$ and $X_n = (Z_n, t_n)$.

a. Verify that the BLUE of θ is $\hat{\theta} = (X_n' X_n)^{-1} X_n' Y_n$ with $\text{Var}(\hat{\theta}) = \sigma^2 (X_n' X_n)^{-1}$.
b. Show that $\text{Var}(\hat{\alpha}) = \sigma^2 \{n - t_n' Z_n (Z_n' Z_n)^{-1} Z_n' t_n\}^{-1}$ and $\text{Var}(\hat{\alpha})$ is minimized when $t_n' Z_n = 0$, which means that t_n is orthogonal to the columns of Z_n ($t_n' \mathbf{1} = 0$ means balance in treatment totals, and $t_n' z_k$ means balance over the kth covariate).

9.8 Consider a linear regression model

$$Y = \mu_A t + \mu_B (1 - t) + \beta z + \epsilon,$$

where $t = 1(0)$ if treatment is $A(B)$, $z = 1(0)$, if a subject is male (female), and $\epsilon \sim N(0, 1)$. Based on a sample of n subjects, let n_{1k} be the number of males assigned to treatment $k = A, B$ and n_{0k} be the number of females assigned to treatment $k = A, B$ such that $n_A = n_{0A} + n_{1A}$, $n_B = n_{0B} + n_{1B}$, and $n = n_A + n_B$.

a. Show that for model (9.18),

$$X'X = \begin{pmatrix} n_A & 0 & n_{1A} \\ 0 & n_B & n_{1B} \\ n_{1A} & n_{1B} & n_1 \end{pmatrix},$$

b. For Atkinson's (1982) approach, we take the directional derivative

$$d(x_{n+1}, \xi_n) = x'_{n+1} M^{-1}(\xi_n) A (A' M^{-1}(\xi_n) A)^{-1} A' M^{-1}(\xi_n) x_{n+1},$$

where $M^{-1}(\xi_n) = n(X'X)^{-1}$, $A = (1, -1, 0)$ and

$$x'_{n+1} = \begin{cases} (1, 0, z), & \text{if treatment} = A; \\ (0, 1, z), & \text{if treatment} = B, \end{cases}$$

where $z = 1$ if $(n + 1)$th subject is male and $z = 0$ if the subject is female. Show that

$$d(x_{n+1}, \xi_n) = \begin{cases} cn_{1B}^2, & \text{if } x'_{n+1} = (1, 0, 1) \ (\text{treatment} = A, \ \text{male}); \\ cn_{1A}^2, & \text{if } x'_{n+1} = (0, 1, 1) \ (\text{treatment} = B, \ \text{male}); \\ cn_{0A}^2, & \text{if } x'_{n+1} = (1, 0, 0) \ (\text{treatment} = A, \ \text{female}); \\ cn_{0B}^2, & \text{if } x'_{n+1} = (0, 1, 0) \ (\text{treatment} = B, \ \text{female}), \end{cases}$$

where c depends on n_{ik}, $i = 0, 1$, $k = A, B$.

c. Hence, show that Atkinson's randomization procedure in this case is:

$$\phi_{n+1} = \begin{cases} \dfrac{n_{1B}^2}{n_{1A}^2 + n_{1B}^2}, & \text{if the } (n + 1)\text{th patient is male;} \\[2ex] \dfrac{n_{0B}^2}{n_{0A}^2 + n_{0B}^2}, & \text{if the } (n + 1)\text{th patient is female.} \end{cases}$$

9.8 REFERENCES

AICKIN, M. (2001). Randomization, balance and the validity and efficiency of design-adaptive allocation methods. *Journal of Statistical Planning and Inference* **94** 97–119.

ATKINSON, A. C. (1982). Optimum biased coin designs for sequential clinical trials with prognostic factors. *Biometrika* **69** 61–67.

ATKINSON, A. C. (1998). Bayesian and other biased-coin designs for sequential clinical trials. *Tatra Mountains Mathematical Publications* **17** 133–139.

ATKINSON, A. C. (1999). Optimum biased-coin designs for sequential treatment allocation with covariate information. *Statistics in Medicine* **18** 1741–1752.

ATKINSON, A. C. AND DONEV, A. N. (1992). *Optimum Experimental Designs*. Clarendon Press, Oxford.

BALL, F. G., SMITH, A. F. M., AND VERDINELLI, I. (1993). Biased coin designs with Bayesian bias. *Journal of Statistical Planning and Inference* **34** 403–421.

BARBÁCHANO, Y. AND COAD, D. S. (2013). Inference following designs which adjust for imbalances in prognostic factors. *Clinical Trials* **10** 540–551.

BARBÁCHANO, Y., COAD, D. S., AND ROBINSON, D. R. (2008). Predictability of designs which adjust for imbalances in prognostic factors. *Journal of Statistical Planning and Inference* **138** 756–767.

BEGG, C. B. AND IGLEWICZ, B. (1980). A treatment allocation procedure for sequential clinical trials. *Biometrics* **36** 81–90.

BEGG, C. B. AND KALISH, L. (1984). Treatment allocation in sequential clinical trials: the logistic model. *Biometrics* **40** 409–420.

BERGER, V. W. (2010). Minimization, by its nature, precludes allocation concealment, and invited selection bias. *Contemporary Clinical Trials* **31** 406–(letter to the editor).

BIRKETT, N. J. (1985). Adaptive allocation in randomized clinical trials. *Controlled Clinical Trials* **6** 146–155.

EFRON, B. (1980). Randomizing and balancing a complicated sequential experiment. In *Biostatistics Casebook* (MILLER, R. G., EFRON, B., BROWN, B. W., AND MOSES, L. E., eds.). John Wiley & Sons, Inc., New York, pp. 19–30.

FORSYTHE, A. B. (1987). Validity and power of tests when groups have been balanced for prognostic factors. *Computational Statistics and Data Analysis* **5** 193–200.

HASEGAWA, T. AND TANGO, T. (2009). Permutation test following covariate-adaptive randomization in randomized clinical trials. *Journal of Biopharmaceutical Statistics* **19** 106–119.

HU, Y. AND HU, F. (2012). Asymptotic properties of covariate-adaptive randomization. *Annals of Statistics* **40** 1794–1815.

HU, F., HU, Y., MA, Z., AND ROSENBERGER, W. F. (2015). Adaptive randomization for balancing over covariates. *WIRES Statistical Computing* **6** 288–303.

HU, F. AND ROSENBERGER, W. F. (2006). *The Theory of Response-Adaptive Randomization in Clinical Trials*. John Wiley & Sons, Inc., New York.

KALISH, L. A. AND HARRINGTON, D. P. (1988). Efficiency of balanced treatment allocation for survival analysis. *Biometrics* **44** 815–821.

KIEFER, J. AND WOLFOWITZ, J. (1960). The equivalence of two extremum problems. *Canadian Journal of Mathematics* **12** 363–366.

KLOTZ, J. H. (1978). Maximum entropy constrained balance randomization for clinical trials. *Biometrics* **34** 283–287.

KUZNETSOVA, O. M. (2010). On the second role of the random element in minimization. *Contemporary Clinical Trials* **31** 587–588.

KUZNETSOVA, O. M. AND TYMOFYEYEV, Y. (2013). Shift in re-randomization distribution with conditional randomization test. *Pharmaceutical Statistics* **12** 82–91.

POCOCK, S. J. AND SIMON, R. (1975). Sequential treatment assignment with balancing for prognostic factors in the controlled clinical trial. *Biometrics* **31** 103–115.

PROSCHAN, M., BRITTAIN, E., AND KAMMERMAN, L. (2011). Minimize the use of minimization with unequal allocation. *Biometrics* **67** 1135–1141.

ROSENBERGER, W. F. AND SVERDLOV, O. (2008). Handling covariates in the design of clinical trials. *Statistical Science* **23** 404–419.

SENN, S. (2004). Controversies concerning randomization and additivity in clinical trials. *Statistics in Medicine* **23** 3729–3753.

SENN, S. (2013). Seven myths of randomisation in clinical trials. *Statistics in Medicine* **32** 1439–1450.

SHAO, J., YU, X., AND ZHONG, B. (2010). A theory for testing hypotheses under covariate-adaptive randomization. *Biometrika* **97** 347–360.

SIBSON, R. (1974). D-optimality and duality. In *Progress in Statistics* (GANI, J., SARKADI, K., AND VINCZE, J., eds.). North-Holland, Amsterdam.

SIGNORINI, D. F., LEUNG, O., SIMES, R. J., BELLER, F., AND GEBSKI, V. J. (1993). Dynamic balanced randomization for clinical trials. *Statistics in Medicine* **12** 2343–2350.

SIMON, R. (1979). Restricted randomization designs in clinical trials. *Biometrics* **35** 503–512.

SIMON, R. AND SIMON, N. R. (2011). Using randomization tests to preserve type I error with response adaptive and covariate adaptive randomization. *Statistics and Probability Letters* **81** 767–772.

STADLER, W. M., LERNER, S. P., GROSHEN, S., STEIN, J. P., SHI, S.-R., RAGHAVAN, D., ESRIG, D., STEINBERG, G., WOOD, D., KLOTZ, L., HALL, C., SKINNER, D. G., AND COTE, R. J. (2011). Phase III study of molecularly targeted adjuvant therapy in locally advanced urothelial cancer of the bladder based on p53 status. *Journal of Clinical Oncology* **29** 3443–3449.

TAVES, D. R. (1974). Minimization: a new method of assigning patients to treatment and control groups. *Clinical Pharmacology and Therapeutics* **15** 443–453.

TAVES, D. R. (2004). Faulty assumptions in Atkinson's criteria for clinical trial design. *Journal of the Royal Statistical Society A* **167** 179–180 (letter to the editor).

TITTERINGTON, D. M. (1983). On constrained balance randomization for clinical trials. *Biometrics* **39** 1083–1086.

WEI, L. J. (1978). An application of an urn model to the design of sequential controlled clinical trials. *Journal of the American Statistical Association* **73** 559–563.

ZELEN, M. (1974). The randomization and stratification of patients to clinical trials. *Journal of Chronic Diseases* **28** 365–375.

10

Response-Adaptive Randomization

10.1 INTRODUCTION

We now revisit a topic mentioned in Chapter 3: the reasons behind equal allocation to the experimental and control therapies. Recall that two reasons were given, namely, power is maximized, and equal allocation reflects the view of equipoise that must exist at the start of the trial. Let us examine these two arguments afresh.

First, power is determined by the information accrued in the clinical trial, and under the traditional concept of statistical information, this is directly related to the variance of the test statistic under the alternative hypothesis. If responses to the treatments have equal variability, power will be maximized under equal allocation. If they do not, power will be maximized using unbalanced allocation, with more patients allocated to the more variable treatment. For binary response problems, variability is directly related to the treatment effectiveness, whereas it will not be related for normal responses. In the latter case, one might have some indication at the beginning of the trial that one treatment is more variable than the other, and sample sizes can begin unbalanced. As the trial progresses, estimates of the variability could be obtained, which would indicate that unequal allocation would result in more power. In the former case, since we are in a state of equipoise (and are essentially operating under the null hypothesis), we would not have any cause to employ unequal allocation, but this again may change as we accrue information on the treatment effect. So, the power issue is considerably more complex than the oft-heard statement "unequal allocation results in a loss of power."

Similarly, should our view of equipoise at the beginning of the trial be fixed throughout the course of the trial, or could we use accruing data to dynamically alter the allocation probabilities to favor the treatment performing best thus far? It would seem that patients would then benefit by having less allocations to an "inferior" treatment (or at least inferior based on the data accrued thus far).

Randomization in Clinical Trials: Theory and Practice, Second Edition.
William F. Rosenberger and John M. Lachin.
© 2016 John Wiley & Sons, Inc. Published 2016 by John Wiley & Sons, Inc.

These are important practical and ethical questions that have been prominent in the literature since the 1950s. So prominent has been the debate that it is surprising that the need for equal allocation is nearly unquestioned in clinical trials of today.

In this chapter, we deal with *response-adaptive randomization*,[1] in which the probability of being assigned to a treatment is changed throughout the trial according to the data that have already accrued. The goal of response-adaptive randomization is to achieve some ethical or statistical objectives that may not be possible by fixing an allocation in advance. These techniques fall under the broad category of *adaptive designs* (distinguished from response-adaptive *randomization*, which refers to *randomized* adaptive designs). Adaptive designs are useful in many disciplines (e.g., Flournoy and Rosenberger, 1995) and have been proposed for clinical trials for many decades. Initial adaptive designs for clinical trials arose from considerations of optimal decision theory, including bandit problems, and of sequential stopping boundaries, and most of these designs were deterministic. We briefly review these designs from a historical perspective. We then discuss techniques for response-adaptive randomization, which affords the protections offered by all randomized experiments.

10.2 HISTORICAL NOTES

Adaptive designs in the clinical trial context were first formulated as solutions to optimal decision-making questions: Which treatment is better? What sample size should be used before determining a "better" treatment to maximize the total number receiving the better treatment? How do we incorporate prior data or accruing data into these decisions? The preliminary ideas can be traced back to Thompson (1933) and Robbins (1952) and led to a flurry of work in the 1960s by Anscombe (1963), Colton (1963), Zelen (1969) and Cornfield, Halperin, and Greenhouse (1969), among others. Perhaps the simplest of these adaptive designs is the *play-the-winner rule* originally explored by Robbins (1952) and later by Zelen (1969), in which a success on one treatment results in the next patient's assignment to the same treatment, and a failure on one treatment results in the next patient's assignment to the opposite treatment.

10.2.1 Roots in bandit problems

Consider a slot machine with two arms and a payoff that is observed immediately. To maximize the total payoff, which arm does one choose to play each time? In the context of clinical trials, the arms are the two treatments, and we desire to optimize some single objective, such as the mean squared error of an estimate of the treatment effect or the expected number of treatment failures. Such optimization problems are called *bandit problems* (cf. Berry and Fristedt, 1985; Gittins, 1989; Hardwick, 1995) and were originally proposed by Thompson (1933) and Robbins (1952).

---

[1] Since the first edition of this book, three new books have been published on response-adaptive randomization, and readers interested in delving further into the details and theory of these methods are referred to Hu and Rosenberger (2006), Atkinson and Biswas (2013), and Baldi Antognini and Giovagnoli (2015).

Ideally, one would like to switch back and forth from one treatment to the other any number of times to obtain the optimal sequence of treatment assignments. This is an extremely difficult problem even in the binary response case, because in order to find the optimal sequence, we have to specify a treatment to be used in each of the 4^n possible paths in a clinical trial with sample size n (i.e., at each allocation, we could have a success on A, failure on A, success on B, failure on B). These problems also involve unknown parameters, and Bayesian and minimax approaches have been employed; see Berry and Fristedt (1985, Chapter 1) for a review of these techniques.

Discrete bandit problems can be solved using *dynamic programming* (Bellman, 1956). In the past, dynamic programming algorithms for even moderate sample sizes were computationally infeasible, but the advent of parallel processing and faster computers has facilitated research into the feasibility and properties of these approach. Much of the early work in developing computational algorithms was done by Hardwick and Stout (1995, 1999).

The difficulty in finding the optimal sequence using dynamic programming led some researchers to find alternative adaptive allocation procedures (e.g., Berry, 1978). Most of these procedures are *myopic strategies*, in which the allocation rule attempts to optimize the treatment assignment for the current patient, by allocating to the treatment that has performed best thus far in the trial. It is well known that myopic strategies are not necessarily globally optimal (Berry and Fristedt, 1985, p.4). Bandit solutions have the advantage that they balance the myopic goal (the patient at hand) with future rewards.

Optimal sequences from bandit solutions are deterministic. There has been very little literature on randomized bandit solutions. Berry and Eick (1995, p. 232) suggest the following:

> Assignment bias can be avoided ... by introducing an unbalanced randomization in which the treatment opposite from the one assigned by the procedure is used with probability sufficiently great to ensure blindness but not so large that the advantage of the adaptive procedure is lost—perhaps this probability can be between $1/10$ and $1/3$.

Recent work on randomized bandits include Yang and Zhu (2002) and Villar, Bowden, and Wason (2015).

10.2.2   Roots in sequential stopping problems

The previous discussion involved a fixed sample size. Others have examined adaptive designs in the context of a random number of patients N, in conjunction with an appropriate stopping boundary. The early papers taking this approach were Chernoff and Roy (1965), Flehinger and Louis (1971), Robbins and Siegmund (1974), Louis (1975), and Hayre (1979), among others.

In the Robbins and Siegmund (1974) model, assume that responses $x_1, \ldots, x_m$ and $y_1, \ldots, y_n$ are realizations of random variables that are independent and identically distributed as $N(\mu_1, 1)$ and $N(\mu_2, 1)$, respectively, and it is desired to test the hypothesis $H_0 : \mu_1 > \mu_2$ versus $H_1 : \mu_1 < \mu_2$. After each response, we observe the test statistic

$$z_{m,n} = \frac{mn}{m+n}(\bar{y}_n - \bar{x}_m).$$

Let $b > 0$ be a constant; we stop the trial as soon as $z_{m,n} \notin (-b, b)$ and declare H_0 is true if $z_{m,n} < -b$ and H_1 is true if $z_{m,n} > b$. Under an appropriate choice of b, we have Wald's sequential probability ratio test with fixed error probabilities α and β. If we wish to minimize the expected number of observations on the treatment with the smaller mean (i.e., the expected number of patients on the inferior treatment), it is logical to assume that equal allocation is preferable when $z_{m,n}$ is close to 0, and when $z_{m,n}$ is close to b or $-b$, most observations should be taken from the x population or the y population, respectively. Robbins and Siegmund propose the following rule. Let $c \geq b$. Having observed $x_1, \ldots, x_m$ and $y_1, \ldots, y_n$, the next observation should be y_{n+1} if

$$\frac{n-m}{m+n} \leq \frac{z_{m,n}}{c};$$

otherwise, the next observation should be x_{m+1}. The authors give some guidelines as to the choice of c and conclude that the error probabilities are essentially independent of the sampling scheme.

Other rules, other response models, and other types of hypotheses are explored by Flehinger and Louis (1971), Louis (1975), and Coad (1991), among others. As in the decision theory approach, most of these approaches to adaptive designs have used nonrandomized allocation rules.

10.2.3   Roots in randomization

Both the bandit and sequential approaches discussed are fully adaptive designs, in that they select future treatments on the basis of all past information about that treatment. However, they have generally been developed for deterministic allocation and, hence, are subject to biases that may be present in nonrandomized studies. In particular, Bather (1995) found that, for both the Robbins and Siegmund procedure and other adaptive designs, "selection bias can have a substantial effect in distorting the results of comparative experiments" (p. 32). Selection bias, as discussed in Chapter 5, is a serious problem for nonrandomized and unmasked, randomized studies. But much of the recent research in adaptive designs has involved fully randomized designs. Response-adaptive randomization alters the allocation probabilities to reflect the current trend of the data, so that patients are assigned to the most "successful" treatment with probability less than 1.

Wei and Durham (1978) were perhaps the first to discuss response-adaptive randomization in their famous *randomized play-the-winner rule* paper. The rule can be described as follows. An urn contains α balls representing treatment A and α balls representing treatment B. A ball is drawn and replaced. If the ball was type $i = A, B$, treatment i is assigned. A success on one treatment results in the addition of β balls representing that treatment, for a positive integer β. A failure on one treatment results in the addition of β balls representing the opposite treatment. Hence, unlike Zelen's model, we skew the probability of assignment to favor the treatment performing "better" (i.e., less failures/more successes), rather than switching deterministically between treatments. This design is usually designated $RPW(\alpha, \beta)$.

Urn models are only one approach to accomplish response adaptive randomization. We discuss these varied approaches in the remaining portions of this chapter. Because this book is about randomization, future discussion will principally focus on fully randomized adaptive designs.

10.3 OPTIMAL ALLOCATION

In this chapter, we will explore (1) response-adaptive randomization that is based on optimal allocation targets, where a specific criterion is optimized based on a population response model, and (2) design-driven response-adaptive randomization, where myopic rules are established, which have intrinsic intuitive motivation and can be completely nonparametric, but are not optimal in a formal sense. Let us begin with determining optimal allocation targets. Because these targets typically depend on unknown parameters of a response distribution, they cannot be implemented in practice without some form of estimation. Hardwick and Stout (1995) review several criteria that one may wish to optimize, including expected number of treatment failures, expected number of successes lost, expected number of patients assigned to the inferior treatment, the total expected sample size, the probability of correct selection, or total expected cost. When the goal is to maximize the experience of individual patients in a clinical trial, the first three criteria are often used. One can argue their relative merits; for example, expected number of treatment failures takes into account the randomness inherent in the response model in that some patients may not benefit from the superior treatment, whereas expected number of patients assigned to the inferior treatment ignores this randomness and focuses on what the scientist can actually control. Of course, the first two criteria, expected failures and expected successes lost, refer to binary response trials where we have a clearly defined "success" and "failure."

These optimal allocation rules are derived under simple homogeneous population models, in which responses of patients assigned to the same treatment are assumed to follow the same distribution. This may be an unreasonable assumption, for example, when there are important covariates that effect response or time trends. But it provides a simple working model to explore properties of response-adaptive randomization. In Chapter 12, we discuss the heterogeneity issue from a practical standpoint.

The general optimization approach we will employ derives from the approach of Jennison and Turnbull (2000) and can be traced back to ideas of Hayre (1979). The idea is to fix the variance of the test statistic to be constant and then to find an optimal allocation ratio R^* from the possible values of $R = n_A/n_B$ according to our particular criterion. In Jennison and Turnbull's approach, let Y_{Ai} arise from a $N(\mu_A, \sigma_A^2)$ distribution and Y_{Bi} arise from a $N(\mu_B, \sigma_B^2)$ distribution, $i = 1, 2, \ldots$, and σ_A^2 and σ_B^2 are known. Then the denominator of the usual Z-test is the square root of the variance of $\bar{Y}_A - \bar{Y}_B$, given by

$$\frac{\sigma_A^2}{n_A} + \frac{\sigma_B^2}{n_B},$$

and we set this equal to a constant, say K. Let $n = n_A + n_B$ be a fixed number of patients in a clinical trial. Then we can write $n_A = Rn/(1 + R)$ and $n_B = n/(1 + R)$, and we obtain

$$n = \frac{\sigma_A^2(1 + R) + \sigma_B^2 R(1 + R)}{KR}. \tag{10.1}$$

Let $\theta = \mu_A - \mu_B$ be the true treatment effect. We wish to find the value of R that minimizes

$$u(\theta)n_A + v(\theta)n_B, \tag{10.2}$$

where u and v are appropriately chosen functions of θ. Because we wish to put more patients on treatment A if $\theta > 0$ and more patients on treatment B if $\theta < 0$, Jennison and Turnbull explore functions where u and v are strictly positive, $u(\theta)$ is decreasing in θ for $\theta < 0$, and $v(\theta)$ is increasing in θ for $\theta > 0$. See Jennison and Turnbull (2000, p. 328) for details on choosing these functions. Substituting (10.1) into (10.2), we obtain

$$\frac{u(\theta)(\sigma_A^2 + \sigma_B^2 R) + v(\theta)\left(\frac{\sigma_A^2}{R} + \sigma_B^2\right)}{K}.$$

Taking the derivative with respect to R and equating to zero, we achieve a minimum at

$$R^* = \frac{\sigma_A}{\sigma_B}\sqrt{\frac{v(\theta)}{u(\theta)}}. \tag{10.3}$$

An interesting case arises. If $u = v = 1$, then we have $R^* = \sigma_A/\sigma_B$, which is simply Neyman allocation (see Problem 2.6), and maximizes the power of the usual Z-test. Note that this formulation also presents an alternative, but equivalent, interpretation. When $u = v = 1$, (10.2) finds the optimal allocation to minimize the total sample size for a fixed variance of the test.

The general formulation with binary response was considered by Hayre and Turnbull (1981) in the context of sequential estimation. Rosenberger *et al.* (2001) also deal with binary responses. If we let the responses on treatment A follow a Bernoulli distribution with parameter p_A and the responses on treatment B follow a Bernoulli distribution with parameter p_B, we can formulate an optimality criterion as in (10.2); however, now the variances depend on p_A and p_B. We also have a dilemma as to which measure of the treatment effect we wish to use. Let $q_A = 1 - p_A$ and $q_B = 1 - p_B$. We could take the simple difference, $\theta = p_A - p_B$, the relative risk of failure, $\theta = q_B/q_A$, or the odds ratio, $\theta = p_A q_B/p_B q_A$. In any event, if we wish to minimize the expected number of treatment failures, (10.2) can be written with $u(\theta) = q_A$ and $v(\theta) = q_B$. The simple difference measure is analogous to the difference of means in the normal case above, and hence, we obtain

$$R^* = \sqrt{\frac{p_A q_A}{p_B q_B}}\sqrt{\frac{q_B}{q_A}} = \sqrt{\frac{p_A}{p_B}}, \tag{10.4}$$

by (10.3). This differs from Neyman allocation, given by

$$R^* = \sqrt{\frac{p_A q_A}{p_B q_B}}. \tag{10.5}$$

If we use the other measures, we obtain different allocations. Consider the log relative risk measure $\log (q_B/q_A)$. We can use the delta method to write the asymptotic variance as

$$\frac{p_A}{n_A q_A} + \frac{p_B}{n_B q_B}.$$

Substituting $n_A = Rn/(1 + R)$ and $n_B = n/(1 + R)$ and equating to K, we obtain

$$n = \frac{p_A(1 + R)}{q_A RK} + \frac{p_B(1 + R)}{q_B K}.$$

Then our optimization criterion becomes finding the value of R to minimize

$$n\frac{q_A R + q_B}{1 + R} = \frac{p_B q_A^2 R^2 + (p_A + p_B)q_A q_B R + p_A q_B^2}{q_A q_B RK}.$$

Taking the derivative with respect to R and equating to zero, we obtain

$$R^* = \sqrt{\frac{p_A}{p_B}\frac{q_B}{q_A}}.$$

Table 10.1 gives the asymptotic variances and optimal allocation for the three types of measures.

The selection of appropriate measure would normally be dictated by the choice of test statistic. The most common is the simple Z test based on the simple difference; see Lachin (2000, Problem 2.9) for asymptotically equivalent tests based on smooth functions of p_A and p_B. Rosenberger et al. (2001) discuss tests based on the pooled versus the separate variance estimators.

Table 10.1   *Asymptotic variances and optimal allocation for minimizing expected number of failures at a fixed variance, for three measures of the treatment effect from binary response trials.*

Measure	Asymptotic Variance	R^*
Simple difference	$\dfrac{p_A q_A}{n_A} + \dfrac{p_B q_B}{n_B}$	$\sqrt{\dfrac{p_A}{p_B}}$
Log relative risk	$\dfrac{p_A}{n_A q_A} + \dfrac{p_B}{n_B q_B}$	$\sqrt{\dfrac{p_A}{p_B} \dfrac{q_B}{q_A}}$
Log odds ratio	$\dfrac{1}{n_A p_A} + \dfrac{1}{n_A q_A} + \dfrac{1}{n_B p_B} + \dfrac{1}{n_B q_B}$	$\sqrt{\dfrac{p_B}{p_A} \dfrac{q_B}{q_A}}$

10.4   RESPONSE-ADAPTIVE RANDOMIZATION TO TARGET R*

Since the optimal allocation involves unknown parameters of the population model, we cannot implement it in practice, unless we implement it under the null hypothesis, resulting in equal allocation, or under some other "best guess" of the parameter values. In this section, we discuss three sequential methods, the *sequential maximum likelihood procedure*, the *doubly adaptive biased coin design*, and the efficient randomized-adaptive design (ERADE).

10.4.1   Sequential maximum likelihood procedure

As the trial progresses, perhaps the most logical approach would be to substitute current values of parameter estimates for the unknown parameters; that is, if $R^*(\theta)$ is a function of an unknown parameter θ, after $j - 1$ patients, substitute $\hat{\theta}(j - 1)$ for θ. Then we can impose the following allocation rule. Let $\mathcal{F}_n = (T_1, \dots, T_n, Y_1, \dots, Y_n)$, where $T_1, \dots, T_n$ assume the value 1 if treatment A and 0 if treatment B, and $Y_1, \dots, Y_n$ are the first n responses to treatment. Then, using similar notation as in Chapter 3, the allocation rule is given by

$$E(T_j | \mathcal{F}_{j-1}) = \frac{R^*(\hat{\theta}(j - 1))}{1 + R^*(\hat{\theta}(j - 1))}. \tag{10.6}$$

While the $\hat{\theta}$ can be any estimator, it is usual to employ the maximum likelihood estimator under the assumed population model. Then the allocation rule in (10.6) is called the *sequential maximum likelihood procedure*.

For example, let us assume the simple binomial model where $Y_i = 1$ if there is a treatment success and $Y_i = 0$ if there is a treatment failure, $i = 1, \dots, n$. The allocation rule for minimizing the expected number of treatment failures under binary response with the simple difference measure, from (10.4), is given by

$$E(T_j | \mathcal{F}_{j-1}) = \frac{\sqrt{\hat{p}_A(j - 1)}}{\sqrt{\hat{p}_A(j - 1)} + \sqrt{\hat{p}_B(j - 1)}},$$

where

$$\hat{p}_A(j-1) = \frac{\sum_{i=1}^{j-1} T_i Y_i}{\sum_{i=1}^{j-1} T_i} \text{ and } \hat{p}_B(j-1) = \frac{\sum_{i=1}^{j-1}(1-T_i)Y_i}{\sum_{i=1}^{j-1}(1-T_i)}.$$

Similarly, we can define a sequential maximum likelihood procedure for Neyman allocation, from (10.5), using

$$E(T_j|\mathcal{F}_{j-1}) = \frac{\sqrt{\hat{p}_A(j-1)\hat{q}_A(j-1)}}{\sqrt{\hat{p}_A(j-1)\hat{q}_A(j-1)} + \sqrt{\hat{p}_B(j-1)\hat{q}_B(j-1)}},$$

where $\hat{q}_A(j-1) = 1 - \hat{p}_A(j-1)$ and $\hat{q}_B(j-1) = 1 - \hat{p}_B(j-1)$. Properties of the sequential maximum likelihood procedure targeting Neyman allocation are explored by Melfi and Page (1995) and Melfi, Page, and Geraldes (2001). Properties of the sequential maximum likelihood procedure targeting (10.4) are explored in Rosenberger *et al.* (2001).

In the latter paper, a simulation was conducted to compare equal allocation, sequential maximum likelihood procedure targeting Neyman allocation, and sequential maximum likelihood procedure targeting (10.4). Results are given in Table 10.2 (the sample sizes were selected to give approximately 90 percent power for the usual Z-test under equal allocation). One can see that the Neyman allocation places too many patients on the inferior treatment for large values of p_A and p_B (see also Problem 10.4). Note that the allocation rule for minimizing expected treatment failures does put fewer patients on the inferior treatment, it is more variable than equal allocation. In general, this will be the case with sequential maximum likelihood procedures, and the variability is induced by the correlation among the treatment assignments. We will reflect on this more when we discuss power in the next chapter.

In what way does the sequential maximum likelihood procedure target the optimal allocation? It seems intuitively reasonable, since maximum likelihood estimators are

Table 10.2 *Simulated values of expected allocation proportions, $E(N_A(n)/n)$ (standard deviation), for the sequential maximum likelihood procedure targeting (10.4) (A), the sequential maximum likelihood procedure targeting Neyman allocation (N), and equal allocation (E), 5000 replications.*

p_A	p_B	n	A	N	E
0.1	0.2	526	0.42 (0.04)	0.43 (0.04)	0.50 (0.02)
0.1	0.3	162	0.39 (0.06)	0.42 (0.05)	0.50 (0.04)
0.1	0.4	82	0.38 (0.07)	0.42 (0.06)	0.50 (0.05)
0.4	0.6	254	0.45 (0.04)	0.50 (0.03)	0.50 (0.03)
0.6	0.9	82	0.45 (0.06)	0.58 (0.06)	0.50 (0.05)
0.7	0.9	162	0.47 (0.04)	0.58 (0.05)	0.50 (0.04)
0.8	0.9	526	0.48 (0.02)	0.57 (0.04)	0.50 (0.02)

Rosenberger *et al.* (2001, p. 911). Reproduced with permission of John Wiley and Sons.

typically consistent, to assume that $R^*(\hat{\theta}(n)) \to R^*(\theta)$ and that

$$\lim_{n \to \infty} \frac{N_A(n)}{n} = \frac{R^*(\theta)}{1 + R^*(\theta)}, \tag{10.7}$$

and hence, we attain optimal allocation in the limit. However, we no longer have independent data because of the response-adaptive randomization, and the proof is considerably more complicated. This issue is addressed in detail in Hu and Rosenberger (2006, Chapter 5). It turns out that, under very mild conditions, (10.7) is true, and hence, the sequential maximum likelihood procedure is asymptotically optimal. The reader is referred to Hu and Rosenberger (2006) for mathematical details.

For the difference of normal means, Jennison and Turnbull (2000) propose a group sequential adaptive design. Suppose there are K interim inspections of the data. At stage $k, k = 1, \dots, K$, the optimal allocation ratio is determined from (10.3) by substituting the current estimates of θ, $\hat{\theta}(k-1)$, from the previous stages into $u(\theta)$ and $v(\theta)$. Then the next group will have size $n_{Ak} + n_{Bk} = n_k$, where n_{Ak} and n_{Bk} are determined by the estimated optimal allocation proportions (to an integer approximation) and n_k is a function of the amount of information accrued. This is a deterministic rule; Jennison and Turnbull mention that one could set a minimum sample size n^* for both arms to preserve at least some "randomization" in the treatment allocation and maintain an element of masking. One could alternatively establish a sequential maximum likelihood procedure that randomizes patients one by one using the estimated optimal allocation as the allocation probability. In this case, one must estimate both the mean and variance, unless one assumes known variances, as do Jennison and Turnbull (2000).

10.4.2 Doubly adaptive biased coin design

Eisele (1994) and Eisele and Woodroofe (1995) propose a more complicated design to achieve the desired allocation proportion R^*. They refer to this design as the *doubly adaptive biased coin design*. Let t be a function from $[0, 1]^2$ to $[0, 1]$ such that the following four conditions hold: (i) t is jointly continuous; (ii) $t(a, a) = a$, (iii) $t(a, b)$ is strictly decreasing in a and strictly increasing in b; and (iv) t has bounded derivatives in both arguments. Let $\rho(\theta) = R^*(\theta)/(1 + R^*(\theta))$. The function t will represent a measure of the difference between $N_A(j)/j$ and $\rho(\hat{\theta}(j))$. Then we allocate to treatment A with probability

$$E(T_j | \mathcal{F}_{j-1}) = t \left(\frac{N_A(j-1)}{j-1}, \rho(\hat{\theta}(j-1)) \right).$$

The properties of this design will depend largely on the function t used. Eisele and Woodroofe (1995) show that (10.7) holds, but under somewhat restrictive conditions. Melfi, Page, and Geraldes (2001) point out that the example in the last section of Eisele (1994) does not satisfy the requisite conditions, so one must choose t carefully.

One particularly interesting function is analogous to Smith's restricted randomization procedure in (3.25), and was developed by Hu and Zhang (2004). For $\gamma \geq 0$:

$$t\left(\frac{N_A(j-1)}{j-1}, \rho(\hat{\theta}(j-1))\right) \tag{10.8}$$

$$= \frac{\rho(\hat{\theta}(j-1))\left(\frac{\rho(\hat{\theta}(j-1))}{\frac{N_A(j-1)}{j-1}}\right)^{\gamma}}{\rho(\hat{\theta}(j-1))\left(\frac{\rho(\hat{\theta}(j-1))}{\frac{N_A(j-1)}{j-1}}\right)^{\gamma} + (1 - \rho(\hat{\theta}(j-1)))\left(\frac{1-\rho(\hat{\theta}(j-1))}{1-\frac{N_A(j-1)}{j-1}}\right)^{\gamma}}; \tag{10.9}$$

$$t(0, \rho(\hat{\theta}(j-1))) = 1;$$

$$t(1, \rho(\hat{\theta}(j-1))) = 0.$$

When $\gamma = 0$, the equation reduces to

$$t\left(\frac{N_A(j-1)}{j-1}, \rho(\hat{\theta}(j-1))\right) = \rho(\hat{\theta}(j-1)),$$

which is the sequential maximum likelihood procedure. Like the parameter ρ in Smith's procedure, increasing γ reduces the variability of the procedure. At the limit, $\gamma \to \infty$, the procedure is deterministic, unless $\rho(\hat{\theta}(j-1)) = N_A(j-1)/(j-1)$. One can tune the procedure by setting γ to reflect the degree of randomness desired. The sequential maximum likelihood procedure is too variable, and simulations have shown that $\gamma = 2$ is a reasonable choice. Tymofyeyev, Rosenberger, and Hu (2007) describe a generalization of Hu and Zhang's function for $K > 2$ treatments.

Zhang and Rosenberger (2006) show how to use the doubly adaptive biased coin design to target optimal allocations when responses are continuous. For normally distributed responses, they use the expected mean total response as the outcome and attempt to minimize it for fixed power. This is a continuous analog to the allocations of Rosenberger *et al.* (2001). They assume that a small response is desirable and minimize the total expected response from all patients. They solve the following optimization problem:

$$\begin{cases} \min_{n_A/n_B} & \mu_A n_A + \mu_B n_B \\ \text{s.t.} & \dfrac{\sigma_A^2}{n_A} + \dfrac{\sigma_B^2}{n_B} = K, \end{cases} \tag{10.10}$$

where K is some constant. This yields

$$\rho(\theta) = \frac{\sqrt{\mu_B}\sigma_A}{\sqrt{\mu_B}\sigma_A + \sqrt{\mu_A}\sigma_B} \tag{10.11}$$

(Problem 10.4). When $\mu_A < \mu_B$, there are possible values of σ_A and σ_B, which make n_A/n_B less than $1/2$. While this may maximize power for fixed expected treatment failures, it is not appropriate to allocate more patients to the inferior treatment. Define $R^* = \sigma_A \sqrt{\mu_B}/\sigma_B \sqrt{\mu_A}$ and

$$
s = \begin{cases} 1 & \text{if } (\mu_A < \mu_B \text{ and } R^* > 1) \text{ or } (\mu_A > \mu_B \text{ and } R^* < 1), \\ 0 & \text{otherwise.} \end{cases}
$$

They modify the allocation as follows.

$$
\rho(\theta) = \begin{cases} \dfrac{\sigma_A \sqrt{\mu_B}}{\sigma_A \sqrt{\mu_B} + \sigma_B \sqrt{\mu_A}} & \text{if } s = 1, \\ \dfrac{1}{2} & \text{otherwise.} \end{cases} \tag{10.12}
$$

They then use the function in (10.9) with ρ defined in (10.12).

Zhang and Rosenberger (2006) remark that when a large response is desirable, one cannot obtain the optimal allocation by changing *min* to *max* in (10.10). However, one can minimize $n_A/\mu_A + n_B/\mu_B$ and attain a valid optimal allocation procedure, although this does not have a natural interpretation as minimizing $n_A\mu_A + n_B\mu_B$. Biswas and Mandal (2004) avoid this problem by minimizing

$$
n_A \Phi \left(\frac{\mu_A - c}{\sigma_A} \right) + n_B \Phi \left(\frac{\mu_B - c}{\sigma_B} \right)
$$

instead of $n_A\mu_A + n_B\mu_B$ in (10.10), where c is a constant and $\Phi(\cdot)$ is the cumulative density function of standard normal distribution. This amounts to minimizing the total number of patients with response greater than c. The corresponding allocation rule is

$$
\rho(\theta) = \frac{\sqrt{\Phi \left(\frac{\mu_B - c}{\sigma_B} \right)} \sigma_A}{\sqrt{\Phi \left(\frac{\mu_B - c}{\sigma_B} \right)} \sigma_A + \sqrt{\Phi \left(\frac{\mu_A - c}{\sigma_A} \right)} \sigma_B}.
$$

In the survival setting, Zhang and Rosenberger (2007) derive optimal allocation for parametric Weibull and exponential models with censoring.

10.4.3   Example

In the appendix, SAS code is presented to simulate properties of the doubly adaptive biased coin design. In this simulation, $n = 200$ and we have binary responses with underlying success probabilities $p_A = 0.70$ and $p_B = 0.25$. We use the allocation function in equation (10.4) and Hu and Zhang's function in (10.9) with $\gamma = 2$. We begin the procedure with two permuted blocks of size 6 to accrue some data on S_A and S_B and to ensure N_A and N_B are nonzero. In some cases, the computation of

$\hat{\rho}(\theta)$ still leads to division by zero. One typical technique to deal with this is to add 1 to S_A and S_B and 2 to N_A and N_B. The user specifies the seeds used in the random number generator. We conduct 100,000 replications.

The results show that the expected number of failures is 93.8 (S.D. = 6.2) and the expected proportion of patients assigned to treatment A is 0.62 (S.D. = 0.10). The latter result matches well with asymptotic theory, because we would expect $E(N_A/n)$ to be close to the true value of $\rho(\theta) = \sqrt{0.75}/(\sqrt{0.75} + \sqrt{0.30}) = 0.626$. For the expected number of failures, we see that equal allocation would yield, on average, $(100 \times 0.30) + (100 \times 0.75) = 105$. Consequently, 12 fewer patients, on average, would experience a treatment failure.

10.4.4 Efficient randomized-adaptive design

Hu, Zhang, and He (2009) proposed and ERADE based on Efron's biased coin design. The ERADE is analogous to a discretized version of Hu and Zhang's function, given in (10.9). For a parameter $\alpha \in (0, 1)$, it is defined by

$$
\begin{aligned}
E(T_j|\mathcal{F}_{j-1}) &= \alpha\rho(\hat{\theta}(j-1)), & &\text{if} \quad N_A(j-1)/(j-1) > \rho(\hat{\theta}(j-1)), \\
&= 1/2, & &\text{if} \quad N_A(j-1)/(j-1) = \rho(\hat{\theta}(j-1)), \\
&= 1 - \alpha(1 - \rho(\hat{\theta}(j-1))), & &\text{if} \quad N_A(j-1)/(j-1) < \rho(\hat{\theta}(j-1)).
\end{aligned}
$$

Using results of Burman (1996), it is suggested that α be selected between 0.4 and 0.7. The extension of the ERADE to $K > 2$ treatments is not obvious and has the same issues as those described for Efron's biased coin design in Section 3.11.

10.5 URN MODELS

The preceding optimal allocation rules are based on parametric models, and response-adaptive randomization that targets the optimal allocation is based on maximum likelihood estimates. A *design-driven* approach to the problem has been developed on an independent track from the optimal allocation approach. The basic idea is to use an intuitive rule to adapt the allocation probabilities as each patient enters the trial. While these rules do not have any optimal properties, we can determine limiting properties of the design, which may be attractive in their own right. One approach involves *urn models*, which include the randomized play-the-winner rule as a special case; designs based on urn models are completely nonparametric.

10.5.1 The generalized Friedman's urn model

A typical urn model for response-adaptive randomization is the generalized Friedman's urn model (Athreya and Karlin, 1968). Initially, a vector $Y_1 = (Z_{11}, \ldots, Z_{1K})$ of balls of type $1, \ldots, K$ are placed in an urn. Patients sequentially enter the trial. When a patient is ready to be randomized, a ball is drawn at random and replaced. If it was type i, the ith treatment is assigned. We then wait for a random variable ξ (whose probability distribution depends on i) to be observed. An additional d_{ij} balls are added to the urn of type $j = 1, \ldots , K$, where $d_{ij}(\xi)$ is some function on the sample space of ξ. The algorithm is repeated through n stages.

Let $\mathbf{Z}_n = (Z_{n1}, \ldots , Z_{nK})$ be the urn composition when the nth patient arrives to be randomized. Then the probability that the patient will be randomized to treatment j is given by $Z_{nj}/|\mathbf{Z}_n|$, where $|\mathbf{Z}_n| = \sum_{i=1}^{K} Z_{ni}$.

Let $\mathbf{D}(\xi) = ((d_{ij})), i, j = 1, \ldots , K$. First-order asymptotics for the generalized Friedman's urn are determined by the generating matrix of the urn, given by $\mathbf{H} = E\{\mathbf{D}(\xi)\}$. Under certain regularity conditions ($\mathbf{H}$ is positive regular and $\Pr\{d_{ij} = 0 \forall j\} = 0$ for all i),

$$\lim_{n \to \infty} \frac{N_j(n)}{n} = v_j \text{ almost surely,} \qquad (10.13)$$

$j = 1, \ldots , K$, where $v = (v_1, \ldots , v_K)$ is the normalized (i.e., $\sum_{j=1}^{K} v_j = 1$) *left* eigenvector corresponding to the maximal eigenvalue of $\mathbf{H}$ (Athreya and Karlin, 1967).

The generalized Friedman's urn is a natural design for clinical trials of K treatments. Wei (1979) proposed the following simple example of a rule for K treatments. Suppose ξ is a binary outcome, success or failure, and let $d_{ij} = (K-1)\delta_{ij}$ if success on treatment i, and $d_{ij} = (1 - \delta_{ij})$ if failure on treatment i, where δ_{ij} is the Kronecker delta. Assuming that ξ is immediately observable after the patient is randomized, we have $|\mathbf{Z}_n| = |\mathbf{Z}_1| + (K-1)(n-1)$.

10.5.2 The randomized play-the-winner rule

When $K = 2$, and for $\mathbf{Z}_1 = (\alpha, \alpha)$, we have the randomized play-the-winner rule described in Section 10.2.3. This has been the most-studied urn model in the response-adaptive randomization literature. We can explore some of its properties, under a simple population model by letting p_A and p_B be the probabilities of success on treatments A and B, respectively, and $q_A = 1 - p_A$, $q_B = 1 - q_B$. We can write the matrix $\mathbf{H}$ for the $RPW(0, 1)$ rule as

$$\mathbf{H} = \begin{bmatrix} p_A & q_A \\ q_B & p_B \end{bmatrix}. \qquad (10.14)$$

The maximal eigenvalue of this matrix (since it is stochastic) is 1, and the normalized left eigenvector corresponding to the eigenvalue 1 is $q_B/(q_A + q_B)$. Thus, by (10.13), we obtain

$$\lim_{n \to \infty} \frac{N_A(n)}{n} = \frac{q_B}{q_A + q_B}, \qquad (10.15)$$

almost surely, or that

$$\lim_{n \to \infty} \frac{N_A(n)}{N_B(n)} = \frac{q_B}{q_A},$$

almost surely. Consequently, in the limit, the $RPW(0, 1)$ rule allocates according to the relative risk of failure. While this is not optimal in the sense of Section 10.3, this is an intuitively appealing limit.

While finite sample results are intractable for most urn models, the $RPW(\alpha, \beta)$ is simple enough that one can obtain $E(N_A(n))$ and $\text{Var}(N_A(n))$ exactly. A recursion can be developed to determine $E(N_A(n))$ as follows. Again, let $Y_1, \ldots, Y_n$ be the responses of the n patients, where $Y_j = 1$ if success and 0 if failure. Let $T_1, \ldots, T_n$ be the treatment assignments, $T_j = 1$ if A, and 0 if B, and define $F_n = (Y_1, \ldots, Y_n, T_1, \ldots, T_n)$. Note that

$$E(T_j Y_j) = \Pr(T_j = 1, Y_j = 1)$$
$$= \Pr(Y_j = 1 | T_j = 1) \Pr(T_j = 1)$$
$$= p_A E(T_j). \tag{10.16}$$

Also, we can show that

$$E(Y_j) = p_B + (p_A - p_B)E(T_j) \tag{10.17}$$

(Problem 10.7). The probability of selecting a type A ball is simply the proportion of balls of type A in the urn. After $n - 1$ patients, we have $2\alpha + \beta(n - 1)$ total balls in the urn, and the number of type A balls is the starting number, α, plus the number of successes on A, plus thenumber of failures on B. Hence, we have

$$E(T_n | F_{n-1}) = \frac{\alpha + \beta \sum_{i=1}^{n-1} T_i Y_i + \beta \sum_{i=1}^{n-1}(1 - T_i)(1 - Y_i)}{2\alpha + \beta(n - 1)}. \tag{10.18}$$

Taking another expectation, we obtain unconditionally

$$E(T_n) = EE(T_n | F_{n-1})$$
$$= \frac{\alpha - \beta \sum_{i=1}^{n-1} E(T_i) - \beta \sum_{i=1}^{n-1} E(Y_i) + 2\beta \sum_{i=1}^{n-1} E(Y_i T_i)}{2\alpha + \beta(n - 1)},$$
$$= \frac{\alpha + \beta(n - 1)q_B}{2\alpha + \beta(n - 1)} + \frac{\beta(p_A - q_B)}{2\alpha + \beta(n - 1)} \sum_{i=1}^{n-1} E(T_i),$$

using (10.16) and (10.17). Noting that

$$E(N_A(n)) = \sum_{i=1}^{n} E(T_i) = E(T_n) + \sum_{i=1}^{n-1} E(T_i),$$

we can write this as

$$E(N_A(n)) = \frac{\alpha + \beta(n - 1)q_B}{2\alpha + \beta(n - 1)} + \left(1 + \frac{\beta(p_A - q_B)}{2\alpha + \beta(n - 1)}\right) \sum_{i=1}^{n-1} E(T_i).$$

We see that we have a recursion of the form

$$C_n = A_n + B_n C_{n-1}$$

with $A_1 = C_1 = 1/2$. The solution to this recursion is

$$C_n = \sum_{i=1}^{n} A_i \prod_{k=i+1}^{n} B_k.$$

Consequently, we have shown that

$$E(N_A(n)) = \sum_{i=1}^{n} \frac{\alpha + \beta(i-1)q_B}{2\alpha + \beta(i-1)} \prod_{k=i+1}^{n} \left(1 + \frac{\beta(p_A - q_B)}{2\alpha + \beta(k-1)} \right)$$

(Rosenberger and Sriram, 1997). The form of $Var(N_A(n))$ is more complicated and is given in Matthews and Rosenberger (1997), requiring at least half a page. They also show that, if $p_A + p_B > 3/2$, the variance depends on the initial urn composition in the limit. This is an undesirable property of urn models for response-adaptive randomization, because the initial urn composition is generally difficult to select in practice, and one would hope that, at least for large samples, the limiting allocation would be invariant to the starting values. We will discuss the selection of α and β in Chapter 12.

Table 10.3 gives $E(N_A(n)/n)$ and S. D.$(N_A(n)/n)$ for the randomized play-the-winner rule with $n = 25$. One can see that the variability is quite large for large values of p_A and p_B. Variability is reduced substantially when $\alpha = 5$, but the adaptive nature of the design is dampened by less extreme allocation to the superior treatment.

Table 10.4 gives the simulated mean allocation proportions (standard deviation) for the $RPW(1, 1)$ rule, which is useful for direct comparison with Table 10.2. It is

Table 10.3 *Exact values of $E(N_A(n)/n)$ (standard deviation) for $n = 25$ for the $RPW(1, 1)$ rule and the $RPW(5, 1)$ rule.*

p_A	p_B	$\alpha = 1$	$\alpha = 5$
0.1	0.3	0.44 (0.09)	0.46 (0.09)
0.1	0.5	0.38 (0.10)	0.42 (0.09)
0.1	0.7	0.29 (0.10)	0.36 (0.10)
0.1	0.9	0.19 (0.10)	0.31 (0.10)
0.3	0.5	0.43 (0.12)	0.45 (0.10)
0.3	0.7	0.35 (0.13)	0.40 (0.11)
0.3	0.9	0.24 (0.13)	0.34 (0.11)
0.5	0.7	0.41 (0.16)	0.45 (0.13)
0.5	0.9	0.30 (0.17)	0.39 (0.13)
0.7	0.9	0.38 (0.21)	0.44 (0.15)

Rosenberger (1999, p. 334), Reproduced with permission of Elsevier.

Table 10.4 *Simulated values of expected allocation proportions, $E(N_A(n)/n)$ (standard deviation), for the RPW(1, 1) procedure, 5000 replications.*

p_A	p_B	n	RPW(1, 1)
0.1	0.2	526	0.47 (0.02)
0.1	0.3	162	0.44 (0.04)
0.1	0.4	82	0.40 (0.05)
0.4	0.6	254	0.40 (0.05)
0.6	0.9	82	0.29 (0.13)
0.7	0.9	162	0.32 (0.13)
0.8	0.9	526	0.38 (0.12)

Rosenberger *et al.* (2001, p. 911), Reproduced with permission of John Wiley and Sons.

clear that the rule is more variable than the sequential maximum likelihood procedure for large values of p_A and p_B, but is less variable (but also more conservative) for small values of p_A and p_B.

10.5.3 Designs to target any allocation

A deficiency of the generalized Friedman's urn model is that the urn targets a fixed allocation v, determined by the left eigenvector corresponding to the maximum eigenvalue of H, which may not be optimal for considerations in the clinical trial. For example, as we have seen, the randomized play-the-winner rule targets the relative risk measure, but this is not optimal for considerations of power. Zhang, Hu, and Cheung (2009) describe a *sequential estimation-adjusted urn* that can target any allocation. The procedure can best be illustrated for $K = 2$ treatments. First, the number of balls added to urn at each stage must be allowed to take noninteger values. Assume the total number of balls added at each stage is fixed, say γ. We wish to find a matrix H such that the eigenvector equation

$$vH = \lambda_{\max}v$$

holds, where $\lambda_{\max}$ is the maximal eigenvalue of H. Then writing H in the form

$$H = \begin{bmatrix} \gamma - a & a \\ b & \gamma - b \end{bmatrix},$$

we solve for a and b to obtain the desired v.

For example, if we wish to target $v = (\sqrt{p_A}/(\sqrt{p_A} + \sqrt{p_B}), \sqrt{p_B}/(\sqrt{p_A} + \sqrt{p_B}))$ (see Section 10.3), then $a = \sqrt{p_B}, b = \sqrt{p_A}$, and $\gamma = \sqrt{p_A} + \sqrt{p_B}$. The rule would then be implemented as follows: after the jth patient is randomized, no matter what the response of the jth patient is, we add $\sqrt{\hat{p}_A(j-1)}$ type A balls to the urn and $\sqrt{\hat{p}_B(j-1)}$ type B balls to the urn. Zhang, Hu, and Cheung (2009) show that the proportion of patients assigned to each treatment converges almost surely to the target allocation and is asymptotically normal. They also demonstrate that the sequential estimation-adjusted urn can target the same allocation as the randomized play-the-winner rule, but with less variability.

10.5.4 Ternary urn models

Another class of urn models for clinical trials of two treatment is the *ternary urn*, described by Ivanova and Flournoy (2001). Suppose there are three possible outcomes, A, B, and C. A ball is drawn and replaced, and the appropriate treatment assigned. If the patient's outcome is A at treatment $i, i = 1, \dots , K$, a type i ball is added to the urn; if the outcome is B, nothing is done; if the outcome is C, a type i ball is removed. When this model is reduced to only two outcomes, we have three types of urn models. For two outcomes A and B, we have Durham and Yu's (1990) urn, where a ball is added if there is a success and the urn remains unchanged if there is a failure. For outcomes A and C, we have the *birth and death urn* of Ivanova *et al.* (2000), where a ball is added if there is a success and a ball is removed if there is a failure. Finally, if we have outcomes B and C, we have the *drop-the-loser rule* (e.g., Ivanova, 2003), in which a ball is removed if there is a failure and the urn remains the same if there is a success.

It can be shown that the limiting composition of Durham and Yu's urn will contain only balls representing the best treatment (Durham, Flournoy, and Li, 1998), provided there is a single best treatment (i.e., for success probabilities $p_1, \dots , p_K$, there exists a unique maximum $p^* \in (p_1, \dots , p_K)$). The almost sure limiting allocation is given by

$$\lim_{n \to \infty} \frac{N_j(n)}{n} = 1, \text{ if } p^* = p_j;$$
$$= 0, \text{ otherwise.} \qquad (10.19)$$

Other authors have modified Durham and Yu's urn to allow noninteger numbers of replacement balls. These urns have been called *randomly reinforced urns* (see Muliere, Paganoni, and Secchi, 2006; Aletti, May, and Secchi, 2009).

The birth and death urn has the following limiting allocation. If $p^* < 1/2$,

$$\lim_{n \to \infty} \frac{N_j(n)}{n} = \frac{\frac{1}{p_j - q_j}}{\sum_{i=1}^{K} \frac{1}{p_i - q_i}}.$$

If $p^* \geq 1/2$, we obtain the same limiting allocation as for Durham and Yu's urn. The limiting allocation of the drop-the-loser rule is given by

$$\lim_{n \to \infty} \frac{N_j(n)}{n} = \frac{\frac{1}{q_j}}{\sum_{i=1}^{K} \frac{1}{q_i}}.$$

For $K = 2$, this is the same limiting allocation as for the randomized play-the-winner rule, given in (10.15).

Urn models that remove balls have a positive probability that certain types of balls will become extinct. In order to eliminate this possibility, Ivanova *et al.* (2000) introduced *immigration balls* that replenish the urn according to a Poisson immigration process. This is equivalent to introducing an additional type of ball in the urn, *immigration balls*. An urn contains balls of $K + 1$ types, representing K treatments, and ball type $K + 1$ representing immigration balls. If ball types 1, ... , K are drawn, the appropriate rule (birth and death or drop-the-loser) is implemented. If an immigration ball is drawn, the ball is returned to the urn along with K additional balls, one of each of types 1, ... , K.

For $K = 2$, Ivanova and Rosenberger (2001) demonstrate that the drop-the-loser rule induces less variability when p_A and p_B are large. Since the randomized play-the-winner rule and drop-the loser rule have identical limiting allocations, but the randomized play-the-winner rule is quite variable for large values of p_A and p_B, it seems reasonable to conclude that the drop-the-loser rule is preferable for highly successful treatments in binary response trials, in terms of variability.

10.5.5 Klein urn

Galbete, Moler, and Plo (2014) describe a *Klein urn model* for response-adaptive randomization. Using the notation of Section 10.5.2, we begin with α balls of each color. A ball is drawn and *not* replaced, and the corresponding treatment is assigned to the patient. If it results in a success, no action is taken. If it results in a failure, a ball of the opposite color is replaced. Then the allocation rule is given by

$$E(T_j|\mathcal{F}_{j-1}) = \frac{\alpha + \sum_{i=1}^{j-1}(1 - T_i)(1 - Y_i) - \sum_{i=1}^{j-1} T_i(1 - Y_i)}{2\alpha}.$$

They demonstrate that the Klein urn has similar properties to the drop-the-loser rule. The Klein urn will be denoted by Klein(α).

10.6 TREATMENT EFFECT MAPPINGS

Rosenberger (1993) introduced the idea of a *treatment effect mapping*, in which allocation probabilities are some function of the current treatment effect. Let $g : \Omega \to [0, 1]$, continuous, such that $g(0) = 1/2$, $g(x) > 1/2$ if $x > 0$, and $g(x) < 1/2$ if $x < 0$. Let $S \in \Omega$ be some measure of the true treatment effect, and let S_j be the observed value of the treatment effect measure after j responses, where $S_j = 0$ if the treatments are equal, $S_j > 0$ if A is better than B, and $S_j < 0$ if A is worse than B. Then we allocate to treatment A with probability

$$E(T_j|\mathcal{F}_{j-1}) = g(S_{j-1}).$$

One would presume that such a procedure would have limiting allocation

$$\lim_{n \to \infty} \frac{N_A(n)}{n} = g(S),$$

but this has not yet been proved formally for general functions g and treatment effect S.

For continuous outcomes, Rosenberger (1993) developed a treatment effect mapping for the linear rank test, using $g(x) = 0.5(1 + x)$, where S is the normalized (centered and scaled) linear rank statistic. Bandyopadhyay and Biswas (2001) used the mapping $g(x) = \Phi(x)$, where Φ is the normal distribution function, and S is the usual two-sample t-test.

For survival outcomes, Rosenberger and Seshaiyer (1997) use the mapping $g(x) = 0.5(1 + x)$ where S is the centered and scaled log-rank test. Yao and Wei (1996) suggest using

$$g(x) = 0.5 + xr, \text{ if } xr \in [-0.4, 0.4];$$
$$= 0.1, \text{ if } xr < -0.4;$$
$$= 0.9, \text{ if } xr > 0.4,$$

where S is the standardized Gehan Wilcoxon test and r is a constant reflecting the degree to which one wishes to adapt the trial.

The intuitive appeal of the treatment effect mapping approach is that we allocate according to the magnitude of the treatment effect thus far in the trial. While this is an intuitively attractive allocation procedure, the limiting allocation is not optimal, and many of these designs have poor statistical power. Because urn models tend to be relevant for binary and multinomial responses, treatment effect mappings have been proposed for more general outcomes, such as continuous outcomes and survival outcomes.

10.7   COVARIATE-ADJUSTED RESPONSE-ADAPTIVE RANDOMIZATION

Hu and Rosenberger (2006) define a covariate-adjusted response-adaptive (CARA) procedure as one for which randomization probabilities depend on the entire history of treatment assignments, responses, and covariates, as well as the current covariate vector of the patient. Then the randomization procedure is defined by $\phi_n = E(T_n | \mathcal{F}_{n-1})$, where $\mathcal{F}_{n-1} = \{T_1, \dots, T_{n-1}, Y_1, \dots, Y_{n-1}, Z_1, \dots, Z_n\}$, where Z_n is the covariate vector of patient n. The goal is similar to that of response-adaptive randomization, with the added consideration of a patient's individual covariate structure. By skewing the probabilities of assignment to the treatment performing better among patients with a similar covariate profile, each patient in the clinical trial may benefit. Such considerations are important in the realm of *personalized medicine*, as the probability of assignment is tailored to depend on the outcomes of patients with similar characteristics. It is also a potential solution to the trade-off among balance, efficiency, and ethics, described in Section 9.4. While seemingly attractive, these designs are very complicated. Also, while in principle one can derive an optimal allocation ratio, it will depend on both the unknown parameters *and* the values of the covariates, which are unknown in advance.

Perhaps the first CARA design was described in Rosenberger, Vidyashankar, and Agarwal (2001) using a treatment effect mapping. They investigated the covariate-adjusted odds ratio of the treatments using a logistic regression model, adjusted for covariates and covariate-by-treatment interactions. Zhang *et al.* (2007) extend this idea to a very general framework that includes all generalized linear models. For a particular covariate vector z, their procedure is designed to target a specific allocation proportion $\rho(\theta, z)$, analogous to the sequential maximum likelihood approach in response-adaptive randomization ignoring covariates. Sverdlov, Rosenberger, and Ryeznik (2013) demonstrate the utility of these procedures using parametric survival models. Asymptotic properties of these procedures are given in Zhang *et al.* (2007) and Hu and Rosenberger (2006, Chapter 9). Properties of CARA designs targeting $\rho(\theta, z)$ can only be evaluated for a fixed covariate vector z.

Atkinson and Biswas (2005) extend the ideas of Section 9.3 to the case of regression models with covariates. Continuing in the context of that section, let $d(k, \theta, z)$ be the directional derivative of the determinant of the information matrix arising from a regression model adjusted for covariate vector z, for treatment $k = A, B$. Then the randomization procedure is given by

$$E(T_j|\mathcal{F}_{j-1}) = \frac{\rho(\hat{\theta}(j-1), z_n)d(A, \hat{\theta}(j-1), z_j)}{\rho(\hat{\theta}(j-1), z_j)d(A, \hat{\theta}(j-1), z_j) + (1 - \rho(\hat{\theta}(j-1), z_j))d(B, \hat{\theta}(j-1), z_j)}.$$

(10.20)

The extension to $K > 2$ treatments is obvious.

Rosenberger and Sverdlov (2008) compare CARA procedures for two treatments and binary responses for four different target allocations with Atkinson and Biswas's procedure in (10.20), by simulation. The simulation evaluates the target allocations at covariate vectors generated as independent Bernoulli, uniform, and normal. They conclude that the nonoptimal target allocation of Rosenberger, Vidyashankar, and Agarwal (2001) skews more patients to the better treatment than the other procedures, but at a considerable loss of power.

10.8 PROBLEMS

10.1 Use the delta method to derive the expressions for the asymptotic variance of the relative risk and odds ratio measures, given in Table 10.1.

10.2 Show that the optimal allocation for the log odds ratio measure using the criterion of Section 10.3 is given by

$$R^* = \sqrt{\frac{p_B \, q_B}{p_A \, q_A}}.$$

10.3 For $p_A = 0.1, 0.5, 0.9$, draw plots superimposing the following allocations across values of p_B:

(i) optimal allocation for the simple difference given in Table 10.1;
(ii) optimal allocation for the log relative risk given in Table 10.1;
(iii) optimal allocation for the log odds ratio given in Table 10.1;
(iv) limiting allocation for the randomized play-the-winner rule, given in (10.15).
Interpret.

10.4 Show that Neyman allocation assigns more patients to the inferior treatment when $p_A > q_B$.

10.5 Show that the solution to the optimization problem in (10.10) is given by (10.11).

10.6 a. Show (10.14).
b. Show that the normalized left eigenvector corresponding to the eigenvalue 1 in (10.14) is given by $(q_B/(q_A + q_B), q_A/(q_A + q_B))$.

10.7 Show (10.17).

10.8 Show that the solution to the recursion

$$C_n = A_n + B_n C_{n-1}$$

with $A_1 = C_1 = 1/2$ is given by

$$C_n = \sum_{i=1}^{n} A_i \prod_{k=i+1}^{n} B_k.$$

10.9 Suppose we have already assigned 9 patients, 5 to A and 4 to B. We have observed success rates of $\hat{p}_A = 3/5$ and $\hat{p}_B = 1/4$. Compute the probability that patient 10 will be assigned to A for the following response-adaptive randomization procedures (Rosenberger and Hu, 2004):

(i) the sequential maximum likelihood procedure targeting $\rho = \sqrt{p_A}/(\sqrt{p_A} + \sqrt{p_B})$;
(ii) the doubly adaptive biased coin design in (10.9) for $\rho = 2$, targeting the allocation in (i);
(iii) RPW(1,1);
(iv) Klein(1).

10.10 Generate a randomization sequence for $n = 50$ for a binary response trial with $p_A = 0.5$ and $p_B = 0.7$ using the following response-adaptive randomization procedures:

(i) the sequential maximum likelihood procedure targeting $\rho = \sqrt{p_A}/(\sqrt{p_A} + \sqrt{p_B})$;
(ii) the doubly adaptive biased coin design in (10.9) for $\gamma = 2$, targeting the allocation in (i);
(iii) the ERADE with $\alpha = 0.5$;
(iv) RPW(1,1);
(v) Klein(1).

10.11 Repeat the simulation in Section 10.4.3 for the ERADE with $\alpha = 0.50$ (same ρ) and the randomized play-the-winner rule. Draw conclusions. What other considerations would you simulate before deciding on an appropriate procedure?

10.9 REFERENCES

ALETTI, G., MAY, C., AND SECCHI, P. (2009). A central limit theorem, and related results, for two-color randomly reinforced urn. *Annals of Applied Probability* **41** 829–844.

ANSCOMBE, F. J. (1963). Sequential medical trials. *Journal of the American Statistical Association* **58** 365–384.

ATHREYA, K. B. AND KARLIN, S. (1967). Limit theorems for the split times of branching processes. *Journal of Mathematics and Mechanics* **17** 257–277.

ATHREYA, K. B. AND KARLIN, S. (1968). Embedding of urn schemes into continuous time Markov branching processes and related limit theorems. *Annals of Mathematical Statistics* **39** 1801–1817.

ATKINSON, A. C. AND BISWAS, A. (2005). Bayesian adaptive biased-coin designs for sequential clinical trials with covariate information. *Biometrics* **61** 118–125.

ATKINSON, A. C. AND BISWAS, A. (2013). *Randomised Response-Adaptive Designs in Clinical Trials*. Chapman and Hall/CRC, Boca Raton, FL.

BALDI ANTOGNINI, A. AND GIOVAGNOLI, A. (2015). *Adaptive Designs for Sequential Treatment Allocation*. Chapman and Hall/CRC, Boca Raton, FL.

BANDYOPADHYAY, U. AND BISWAS, A. (2001). Adaptive designs for normal responses with prognostic factors. *Biometrika* **88** 409–419.

BATHER, J. (1995). Response adaptive allocation and selection bias. In *Adaptive Designs* (FLOURNOY, N. AND ROSENBERGER, W. F., eds.). Institute of Mathematical Statistics, Hayward, CA, pp. 23–35.

BELLMAN, R. (1956). A problem in the sequential design of experiments. *Sankhya A* **16** 221–229.

BERRY, D. A. (1978). Modified two-armed bandit strategies for certain clinical trials. *Journal of the American Statistical Association* **73** 339–345.

BERRY, D. A. AND EICK, S. G. (1995). Adaptive assignment versus balanced randomization in clinical trials: a decision analysis. *Statistics in Medicine* **14** 231–246.

BERRY, D. A. AND FRISTEDT, B. (1985). *Bandit Problems: Sequential Allocation of Experiments*. Chapman and Hall, London.

BISWAS, A. AND MANDAL, S. (2004). Optimal adaptive designs in phase III clinical trials for continuous responses with covariates. In *mODa 7-Advances in Model-Oriented*

Design and Analysis (BUCCHIANICO, A. D., LÄUTER, H., AND WYNN, H. P., eds.). Physica-Verlag, Heidelberg, pp. 51–58.

BURMAN, C. F. (1996). On Sequential Treatment Allocation in Clinical Trials. Sweden: University of Göteborg (doctoral dissertation).

CHERNOFF, H. AND ROY, S. N. (1965). A Bayes sequential sampling inspection plan. *Annals of Mathematical Statistics* **36** 1387–1407.

COAD, D. S. (1991). Sequential tests for an unstable response variable. *Biometrika* **78** 113–121.

COLTON, T. (1963). A model for selecting one of two medical treatments. *Journal of the American Statistical Association* **58** 388–400.

CORNFIELD, J., HALPERIN, M., AND GREENHOUSE, S. W. (1969). An adaptive procedure for sequential clinical trials. *Journal of the American Statistical Association* **64** 759–770.

DURHAM, S. D., FLOURNOY, N., AND LI, W. (1998). Sequential designs for maximizing the probability of a favourable response. *Canadian Journal of Statistics* **3** 479–495.

DURHAM, S. D. AND YU, C. F. (1990). Randomized play-the-leader rules for sequential sampling from two populations. *Probability in the Engineering and Information Sciences* **4** 355–367.

EISELE, J. R. (1994). The doubly adaptive biased coin design for sequential clinical trials. *Journal of Statistical Planning and Inference* **38** 249–261.

EISELE, J. R. AND WOODROOFE, M. B. (1995). Central limit theorems for doubly adaptive biased coin designs. *Annals of Statistics* **23** 234–254.

FLEHINGER, B. J. AND LOUIS, T. A. (1971). Sequential treatment allocation in clinical trials. *Biometrika* **58** 419–426.

FLOURNOY, N. AND ROSENBERGER, W. F., eds. (1995). *Adaptive Designs*. Institute of Mathematical Statistics, Hayward, CA.

GALBETE, A., MOLER, J. A., AND PLO, F. (2014). A response-driven adaptive design based on the Klein urn. *Methodology and Computing in Applied Probability* **16** 731–746.

GITTINS, J. C. (1989). *Multi-Armed Bandit Allocation Indices*. John Wiley & Sons, Ltd, Chichester.

HARDWICK, J. (1995). A modified bandit as an approach to ethical allocation in clinical trials. In *Adaptive Designs* (FLOURNOY, N. AND ROSENBERGER, W. F., eds.). Institute of Mathematical Statistics, Hayward, CA, pp. 223–237.

HARDWICK, J. AND STOUT, Q. F. (1995). Exact computational analysis for adaptive designs. In *Adaptive Designs* (FLOURNOY, N. AND ROSENBERGER, W. F., eds.). Institute of Mathematical Statistics, Hayward, CA, pp. 65–87.

HARDWICK, J. AND STOUT, Q. F. (1999). Using path induction for evaluating sequential allocation procedures. *SIAM Journal of Scientific Computing* **21** 67–87.

HAYRE, L. S. (1979). Two-population sequential tests with three hypotheses. *Biometrika* **66** 465–474.

HAYRE, L. S. AND TURNBULL, B. W. (1981). Estimation of the odds ratio in the two-armed bandit problem. *Biometrika* **68** 661–668.

HU, F. AND ROSENBERGER, W. F. (2006). *The Theory of Response-Adaptive Randomization in Clinical Trials*. John Wiley & Sons, Inc., New York.

HU, F. AND ZHANG, L-X. (2004). Asymptotic properties of doubly adaptive biased coin designs for multi-treatment clinical trials. *Annals of Statistics* **32** 268–301.

HU, F., ZHANG, L.-X., AND HE, X. (2009). Efficient randomized-adaptive designs. *Annals of Statistics* **37** 2543–2560.

IVANOVA, A. V. (2003). A play-the-winner type urn model with reduced variability. *Metrika*, **58**, 1–13.

IVANOVA, A. AND FLOURNOY, N. (2001). A birth and death urn for ternary outcomes: stochastic processes applied to urn models. In *Probability and Statistical Models with Applications* (CHARALAMBIDES, C. A., KOUTRAS, M. V., AND BALAKRISHNAN, N., eds.). Chapman and Hall/CRC, Boca Raton, FL, pp. 583–600.

IVANOVA, A. AND ROSENBERGER, W. F. (2001). Adoptive designs with highly effective treatments. *Drug Information Journal* **35** 1087–1093.

IVANOVA, A. V., ROSENBERGER, W. F., DURHAM, S. D., AND FLOURNOY, N. (2000). A birth and death urn for randomized clinical trials: Asymptotic methods. *Sankhya B* **62** 104–118.

JENNISON, C. AND TURNBULL, B. W. (2000). *Group Sequential Methods with Applications to Clinical Trials*. Chapman and Hall/CRC, Boca Raton, FL.

LACHIN, J. M. (2000). *Biostatistical Methods: The Assessment of Relative Risks*. John Wiley & Sons, Inc., New York.

LOUIS, T. A. (1975). Optimal allocation in sequential tests comparing the means of two Gaussian populations. *Biometrika* **62** 359–369.

MATTHEWS, P. C. AND ROSENBERGER, W. F. (1997). Variance in randomized play-the-winner clinical trials. *Statistics and Probability Letters* **35** 233–240.

MELFI, V. AND PAGE, C. (1995). Variability in adaptive designs for estimation of success probabilities. In *New Developments and Applications in Experimental Design* (FLOURNOY, N., ROSENBERGER, W. F., AND WONG, W. K., eds.). Institute of Mathematical Statistics, Hayward, CA, pp. 106–114.

MELFI, V. F., PAGE, C., AND GERALDES, M. (2001). An adaptive randomized design with application to estimation. *Canadian Journal of Statistics* **29** 107–116.

MULIERE, P., PAGANONI, A. M., AND SECCHI, P. (2006). A randomly reinforced urn. *Journal of Statistical Planning and Inference* **136** 1853–1874.

ROBBINS, H. (1952). Some aspects of the sequential design of experiments. *Bulletin of the American Mathematical Society* **58** 527–535.

ROBBINS, H. AND SIEGMUND, D. O. (1974). Sequential tests involving two populations. *Journal of the American Statistical Association* **69** 132–139.

ROSENBERGER, W. F. (1993). Asymptotic inference with response-adaptive treatment allocation designs. *Annals of Statistics* **21** 2098–2107.

ROSENBERGER, W. F. (1999). Randomized play-the-winner clinical trials: review and recommendations. *Controlled Clinical Trials* **20** 328–342.

ROSENBERGER, W. F. AND HU, F. (2004). Maximizing power and minimizing treatment failures in clinical trials. *Clinical Trials* **1** 141–147.

ROSENBERGER, W. F. AND SESHAIYER, P. (1997). Adaptive survival trials. *Journal of Biopharmaceutical Statistics* **7** 617–624.

ROSENBERGER, W. F. AND SRIRAM, T. N. (1997). Estimation for an adaptive allocation design. *Journal of Statistical Planning and Inference* **59** 309–319.

ROSENBERGER, W. F., STALLARD, N., IVANOVA, A., HARPER, C. N., AND RICKS, M. L. (2001). Optimal adaptive designs for binary response trials. *Biometrics* **57** 909–913.

ROSENBERGER, W. F. AND SVERDLOV, O. (2008). Handling covariates in the design of clinical trials. *Statistical Science* **23** 404–419.

ROSENBERGER, W. F., VIDYASHANKAR, A. N., AND AGARWAL, D. K. (2001). Covariate-adjusted response-adaptive designs for binary response. *Journal of Biopharmaceutical Statistics* **11** 227–236.

SVERDLOV, O., ROSENBERGER, W. F., AND RYEZNIK, Y. (2013). Utility of covariate-adjusted response-adaptive randomization in survival trials. *Statistics in Biopharmaceutical Research* **5** 38–53.

THOMPSON, W. R. (1933). On the likelihood that one unknown probability exceeds another in the view of the evidence of the two samples. *Biometrika* **25** 275–294.

TYMOFYEYEV, Y., ROSENBERGER, W. F., AND HU, F. (2007). Implementing optimal allocation in sequential binary response experiments. *Journal of the American Statistical Association* **102** 224–234.

VILLAR, S. S., BOWDEN, J., AND WASON, J. (2015). Multi-armed bandit models for the optimal design of clinical trials: benefits and challenges. *Statistical Science* **30** 199–215.

WEI, L. J. (1979). The generalized Pólya's urn design for sequential medical trials. *Annals of Statistics* **7** 291–296.

WEI, L. J. AND DURHAM, S. D. (1978). The randomized play-the-winner rule in medical trials. *Journal of the American Statistical Association* **73** 840–843.

YANG, Y. AND ZHU, D. (2002). Randomized allocation with nonparametric estimation for a multi-armed bandit problem with covariates. *Annals of Statistics* **30** 100–121.

YAO, Q. AND WEI, L. J. (1996). Play the winner for phase II/III clinical trials. *Statistics in Medicine* **15** 2413–2423.

ZELEN, M. (1969). Play the winner and the controlled clinical trial. *Journal of the American Statistical Association* **64** 131–146.

ZHANG, L. AND ROSENBERGER, W. F. (2006). Response-adaptive randomization for clinical trials with continuous outcomes. *Biometrics* **62** 562–569.

ZHANG, L. AND ROSENBERGER, W. F. (2007). Response-adaptive randomization for survival trials: the parametric approach. *Journal of the Royal Statistical Society C* **53** 153–165.

ZHANG, L.-X., HU, F., AND CHEUNG, S. H. (2009). Asymptotic theorems of sequential estimation-adjusted urn models. *Annals of Applied Probability* **16** 340–369.

ZHANG, L.-X., HU, F., CHEUNG, S. H., AND CHAN, W. S. (2007). Asymptotic properties of covariate-adjusted response-adaptive designs. *Annals of Statistics* **35** 1166–1182.

10.10   APPENDIX

Here we present the SAS code used in the simulation example in Section 10.4.3. The user specifies the seeds for the random number generation.

```
data dbcd;
 do reps=1 to 100000;
  gamma=2;
  n=200;
  pa=0.70; pb=0.25;
  na=0; nb=0; sa=0; sb=0;
  nablock=0;
  *Start procedure with 2 blocks of size 6;
  do i=1 to 6;
   x=ranuni(1536);
   y=ranuni(6348);
   *Random allocation rule in each block;
   p=(3-nablock)/(7-i);
   if x < p then do;
    na+1; nablock+1;
 if y < pa then sa+1;
   end;
   else do;
```

```
 nb+1;
 if y < pb then sb+1;
   end;
  end;
  nablock=0;
  do i=7 to 12;
   x=ranuni(87657);
   y=ranuni(4563);
   p=(3-nablock)/(13-i);
   if x < p then do;
    na+1; nablock+1;
 if y < pa then sa+1;
   end;
   else do;
 nb+1;
 if y < pb then sb+1;
   end;
  end;
  do i=13 to n;
   pahat=(sa+1)/(na+2); pbhat=(sb+1)/(nb+2);
   *Allocation in equation (10.4);
   rhohat=sqrt(pahat)/(sqrt(pahat)+sqrt(pbhat));
   p=(rhohat*(rhohat/(na/(i-1)))**gamma)/
      ((rhohat*(rhohat/(na/(i-1)))**gamma)
      +((1-rhohat)*((1-rhohat)/(nb/(i-1)))**gamma));
   x=ranuni(8757);
   y=ranuni(456);
   if x < p then do;
    na+1;
    if y < pa then sa+1;
   end;
   else do;
    nb+1;
    if y < pb then sb+1;
   end;
  end;
  failures=n-sa-sb;
  allocprop=na/n;
  output;
end;
proc means mean var;
 var failures allocprop;
run;
```

11

Inference for Response-Adaptive Randomization

11.1 INTRODUCTION

Inference for response-adaptive randomization is very complicated because both the treatment assignments and responses are correlated. This leads to nonstandard problems and new insights into conditioning. We first examine likelihood-based inference and then randomization-based inference. More details on the theory of likelihood-based inference for response-adaptive randomization can be found in Hu and Rosenberger (2006).

11.2 POPULATION-BASED INFERENCE

11.2.1 The likelihood

As in Section 6.2, we can use conditioning arguments to derive the likelihood for a response-adaptive randomization. Let $t^{(j)} = (t_1, \ldots, t_j)$ and $y^{(j)} = (y_1, \ldots, y_j)$ be the realized treatment assignments and responses from patients $1, \ldots, j$, respectively. Let θ be the parameter vector of interest. Unlike the restricted randomization case, here $(t_1, \ldots, t_n)$ depend on θ. However, we have additional data arising from the experiment: the adaptive mechanism. For the urn model, let $z^{(j)} = (z_0, \ldots, z_{j-1})$ be the history of the urn composition, where z_0 is the initial urn composition and z_i is the urn composition after i stages. Then the likelihood of the data after n patients, denoted as

Randomization in Clinical Trials: Theory and Practice, Second Edition.
William F. Rosenberger and John M. Lachin.
© 2016 John Wiley & Sons, Inc. Published 2016 by John Wiley & Sons, Inc.

$\mathcal{L}_n$, is given by

$$
\begin{aligned}
\mathcal{L}_n &= \mathcal{L}(y^{(n)}, t^{(n)}, z^{(n)}; \theta) \\
&= \mathcal{L}(y_n | y^{(n-1)}, t^{(n)}, z^{(n)}; \theta) \mathcal{L}(t_n | y^{(n-1)}, t^{(n-1)}, z^{(n)}; \theta) \\
&\quad \times \mathcal{L}(z_{n-1} | y^{(n-1)}, t^{(n-1)}, z^{(n-1)}; \theta) \mathcal{L}_{n-1}.
\end{aligned}
\tag{11.1}
$$

Since the responses depend only on the treatment assigned and are independent and identically distributed under a population model, we have

$$
\mathcal{L}(y_n | y^{(n-1)}, t^{(n)}, z^{(n)}; \theta) = \mathcal{L}(y_n | t_n; \theta).
\tag{11.2}
$$

Now the treatment assignments will depend only on the current urn composition at the time of assignment. This means that

$$
\mathcal{L}(t_n | y^{(n-1)}, t^{(n-1)}, z^{(n)}; \theta) = \mathcal{L}(t_n | z_{n-1}).
\tag{11.3}
$$

Noting that the urn composition at stage $n - 1$ is completely determined by the urn composition at stage $n - 2$ and the treatment assignment and response of the $n - 1$th patient, we see that

$$
\mathcal{L}(z_{n-1} | y^{(n-1)}, t^{(n-1)}, z^{(n-1)}; \theta) = 1.
\tag{11.4}
$$

Combining (11.1)–(11.4), we obtain

$$
\begin{aligned}
\mathcal{L}_n &= \mathcal{L}(y_n | t_n; \theta) \mathcal{L}(t_n | z_{n-1}) \mathcal{L}_{n-1} \\
&= \prod_{i=1}^{n} \mathcal{L}(y_i | t_i; \theta) \mathcal{L}(t_i | z_{i-1}).
\end{aligned}
$$

Since $\mathcal{L}(t_i | z_{i-1})$ is independent of θ, we have

$$
\mathcal{L}_n \propto \prod_{i=1}^{n} \mathcal{L}(y_i | t_i; \theta)
\tag{11.5}
$$

(Rosenberger, Flournoy, and Durham, 1997).

Note that (11.5) is identical to the likelihood from restricted randomization, in (6.5). However, this only means that the likelihoods look the same. The distribution of the sufficient statistics is quite different in response-adaptive randomization than in restricted randomization, as we shall see.

For the case where there are K treatments, let $\delta_{ji} = 1$ if $t_i = j, j = 1, \ldots, K$, and 0 otherwise. When we have binary responses and $\Pr(Y_i = 1 | T_i = j) = p_j$, we can write the likelihood as

$$
\begin{aligned}
\mathcal{L}_n &= \prod_{i=1}^{n} \prod_{j=1}^{K} [\mathcal{L}(y_i | t_i; \theta)]^{\delta_{ji}} \\
&= \prod_{i=1}^{n} \prod_{j=1}^{K} p_j^{y_i \delta_{ji}} (1 - p_j)^{(1 - y_i)\delta_{ji}} \\
&= \prod_{j=1}^{K} p_j^{s_j(n)} (1 - p_j)^{n_j(n) - s_j(n)},
\end{aligned}
\tag{11.6}
$$

where $s_j(n) = \sum_{i=1}^{n} y_i \delta_{ji}$ and $n_j(n) = \sum_{i=1}^{n} \delta_{ji}$. Let $S_j(n)$ and $N_j(n)$ be the random analogs of $s_j(n)$ and $n_j(n)$, respectively. Note that the maximum likelihood estimator of p_j is $\hat{p}_j = S_j(n)/N_j(n)$.

Under certain regularity conditions, the maximum likelihood estimators are consistent and asymptotically normal. Hu and Rosenberger (2006) derive the asymptotic distribution of the maximum likelihood estimators for a large class of response-adaptive randomization procedures under an exponential family. The basic condition is that

$$
\frac{N_j(n)}{n} \to \rho_j(\theta)
\tag{11.7}
$$

almost surely, $j = 1, \ldots, K$, where $\rho_j(\theta) \in (0, 1)$ is a constant. They also demonstrate how to compute the asymptotic variances of the estimators as the inverse of the Fisher's information, denoted as $I^{-1}(\theta)$. The main result is that, under condition (11.7),

$$
\sqrt{n}(\hat{\theta} - \theta) \to N(\mathbf{0}, I^{-1}(\theta)),
\tag{11.8}
$$

in distribution. For binary response, the form of the asymptotic variance is

$$
I^{-1}(p_A, p_B) = \begin{bmatrix} \dfrac{p_A q_A}{\rho(p_A, p_B)} & 0 \\ 0 & \dfrac{p_B q_B}{1 - \rho(p_A, p_B)} \end{bmatrix}.
\tag{11.9}
$$

For example, with the randomized play-the-winner rule, we know that

$$
\frac{N_A}{n} \to \frac{q_B}{q_A + q_B},
$$

almost surely, so that

$$\sqrt{n}\left(\begin{bmatrix}\hat{p}_A\\\hat{p}_B\end{bmatrix} - \begin{bmatrix}p_A\\p_B\end{bmatrix}\right) \to N\left(0, \begin{bmatrix}\dfrac{p_A q_A(q_A + q_B)}{q_B} & 0\\0 & \dfrac{p_B q_B(q_A + q_B)}{q_A}\end{bmatrix}\right)$$

in distribution.

Note that condition (11.7) does not hold for Durham and Yu's urn (see (10.19)) and randomly reinforced urns described in Section 10.5.4. However, the t-test can be modified to provide an inference procedure for the treatment effect. The theory of inference for this type of urn model is given in May and Flournoy (2009).

The effect of delayed responses on the asymptotic distribution of maximum likelihood estimators is summarized in Section 7.1 of Hu and Rosenberger (2006), based on work of Bai, Hu, and Rosenberger (2002) and Hu and Zhang (2004). Basically, the result in (11.8) holds provided that the delay is not "too long" relative to the patient entry stream. They show that if patient entries arise as a random sample from a uniform distribution, and patient entries arise from an exponential distribution, then delayed responses have no effect on the result in (11.8).

11.2.2 Sufficiency

We can determine the sufficient statistics for θ from the likelihood. Here is where we must carefully distinguish between restricted and response-adaptive randomization. For restricted designs, $N_j(n)$ does not depend on p_j and, hence, is an ancillary statistic. It follows that $S_j(n)$ is a complete sufficient statistic for p_j. By Basu's theorem (Lehmann (1983, p. 46)), this implies that $S_j(n)$ and $N_j(n)$ are independent.

In contrast, for response-adaptive randomization, $N_j(n)$ does carry information about p_j and, therefore, is not ancillary. In fact, the statistics $(S_1(n), \ldots, S_K(n), N_1(n), \ldots, N_{K-1}(n))$ are jointly sufficient for $(p_1, \ldots, p_K)$. This brings up the interesting dilemma that if we condition on $N_j(n)$ when we do inference, we lose extensive information. Thus, response-adaptive randomization requires *unconditional* tests.

11.2.3 Bias of the maximum likelihood estimators

Because of the dependence structure induced by response-adaptive randomization, the maximum likelihood estimators, although they are typically consistent, are biased. Coad and Ivanova (2001) derive the bias factor as follows. Let $S_A(n)$, $S_B(n)$, $N_A(n)$, $N_B(n)$ be the number of successes on A, successes on B, number of patients on A, patients on B, respectively. Let P_{p_A,p_B} be the probability measure on the sequences of treatment responses determined by p_A and p_B and E_{p_A,p_B} be the expectation with respect to that measure. Then dP_{p_A,p_B} is given by

$$dP_{p_A,p_B} = \frac{N_A(n)! N_B(n)!}{S_A(n)! S_B(n)!(N_A(n) - S_A(n))!(N_B(n) - S_B(n))!}$$
$$\times p_A^{S_A(n)}(1 - p_A)^{N_A(n) - S_A(n)} p_B^{S_B(n)}(1 - p_B)^{N_B(n) - S_B(n)}$$

and the first derivative is given by

$$\frac{\partial}{\partial p_i}dP_{p_A,p_B} = \frac{1}{p_i(1-p_i)}(S_i(n) - p_iN_i(n))dP_{p_A,p_B}, i = A, B.$$

We can write

$$E_{p_A,p_B}\left(\frac{1}{N_i(n)}\right) = \int \frac{1}{N_i(n)}dP_{p_A,p_B}, i = A, B.$$

Assuming that E_{p_A,p_B} is continuous in p_i, we may differentiate under the integral sign to obtain

$$\frac{\partial}{\partial p_i}E_{p_A,p_B}\left(\frac{1}{N_i(n)}\right) = \frac{1}{p_i(1-p_i)}\int\left(\frac{1}{N_i(n)}\right)(S_i(n) - p_iN_i(n))dP_{p_A,p_B}$$

$$= \frac{1}{p_i(1-p_i)}E_{p_A,p_B}(\hat{p}_i - p_i). \tag{11.10}$$

Using (11.10), we can write the bias as

$$E_{p_A,p_B}(\hat{p}_i - p_i) = p_i(1-p_i)\frac{\partial}{\partial p_i}E_{p_A,p_B}\left(\frac{1}{N_i(n)}\right). \tag{11.11}$$

Clearly, this is zero if we do not have response-adaptive randomization.

The delta method will allow us to obtain a suitable approximation to (11.11). We can write

$$E_{p_A,p_B}\left(\frac{1}{N_i(n)}\right) \simeq \frac{1}{E_{p_A,p_B}(N_i(n))} + \frac{\mathrm{Var}_{p_A,p_B}(N_i(n))}{[E_{p_A,p_B}(N_i(n))]^3}, i = A, B, \tag{11.12}$$

(Problem 11.1). For specific response-adaptive randomization procedures, if we can compute the variance of $N_i(n)$, we can obtain an approximate bias correction using (11.11) and (11.12).

As an example, consider the randomized play-the-winner rule with two treatments, where $q_A = 1 - p_A$ and $q_B = 1 - p_B$. It is known that

$$\frac{E(N_A(n))}{n} \to \frac{q_B}{q_A + q_B}$$

and that, when $p_A + p_B < 3/2$,

$$\frac{\mathrm{Var}(N_A(n))}{n} \to \frac{q_Aq_B(1 + 2p_A + 2p_B)}{(q_A + q_B)^2(3 - 2p_A - 2p_B)}$$

(Matthews and Rosenberger, 1997). Then we can obtain the following approximation for the bias:

$$E_{p_A,p_B}(\hat{p}_A - p_A) \simeq p_Aq_A\frac{\partial}{\partial p_A}\left[\frac{q_A + q_B}{nq_B} + \frac{q_A(q_A + q_B)(1 + 2p_A + 2p_B)}{n^2q_B^2(3 - 2p_A - 2p_B)}\right].$$

If we ignore the term of order $O(n^{-2})$, we obtain

$$E_{p_A, p_B}(\hat{p}_A - p_A) = -\frac{p_A q_A}{n q_B} + o(n^{-1}).$$

This correction is reasonably accurate for small sample sizes. Similar bias corrections are given for other urn designs in Coad and Ivanova (2001).

11.2.4   Confidence interval procedures

Confidence interval procedures have been proposed for response-adaptive randomization. These include exact binomial confidence intervals for the randomized play-the-winner rule and a bootstrap procedure for a general response-adaptive randomization procedure of K treatments with binary responses. Coad and Woodroofe (1997) and Coad and Govindarajulu (2000) construct confidence intervals following sequential adaptive designs for censored survival data and binary responses, respectively.

We outline the basic procedure for the computation of exact confidence intervals. Wei *et al.* (1990) derived exact binomial confidence intervals for difference measures of p_A and p_B, such as the simple difference $\Delta = p_A - p_B$ following a clinical trial using the randomized play-the-winner rule. The exact distribution will depend on Δ and a nuisance parameter, p_B. One popular approach for dealing with a nuisance parameter is to condition on a sufficient statistic. Wei, *et al.* chose to maximize over the possible values of p_B. Let S_A, S_B, N_A, N_B be the number of successes on treatments A and B and the numbers of patients on A and B, respectively, with realizations s_A, s_B, n_A, n_B, and let $\hat{\Delta} = s_A/n_A - s_B/n_B$. Then if $n_A, n_B > 0$, the exact unconditional confidence interval $(\underline{\Delta}, \overline{\Delta})$ can be computed according to the formulas

$$\underline{\Delta} = \inf_{-1 \leq \Delta \leq 1} \left\{ \Delta : \left[\max_{p_B} \Pr\left(\frac{S_A}{N_A} - \frac{S_B}{N_B} \geq \hat{\Delta}, N_A > 0, N_B > 0 \right) \right] \geq \alpha_1 \right\},$$

$$\overline{\Delta} = \sup_{-1 \leq \Delta \leq 1} \left\{ \Delta : \left[\max_{p_B} \Pr\left(\frac{S_A}{N_A} - \frac{S_B}{N_B} \geq \hat{\Delta}, N_A > 0, N_B > 0 \right) \right] \geq \alpha_2 \right\},$$

for fixed constants α_1 and α_2. These confidence intervals can be computationally-intensive and rely on the networking algorithm approach. Wei, *et al.* compare their unconditional confidence interval to the conditional confidence interval and found that the unconditional intervals tend to be shorter and more efficient than the conditional counterparts for the randomized play-the-winner rule.

Rosenberger and Hu (1999) derived bootstrap confidence intervals following a general response-adaptive randomization procedure of K treatments, using a simple rank ordering. The algorithm is as follows:

1. Obtain the observed data, $\hat{p} = (\hat{p}_1, \dots, \hat{p}_K)$ and $N = (N_1, \dots, N_K)$, the vector of observed success proportions and sample sizes.

2. Simulate the adaptive allocation rule B times, using $\hat{p}$ as the underlying response probabilities, obtaining B sequences of treatment assignments and responses.

3. Compute $\hat{p}_1^*, \ldots, \hat{p}_B^*$ and $N_1^*, \ldots, N_B^*$ from the simulations. These are the bootstrap estimates of the response probabilities and sample sizes.

4. Order $\hat{p}_i^{*1}, \ldots, \hat{p}_i^{*B}$, for $i = 1, \ldots, K$ as $\hat{p}_i^{*(1)}, \ldots, \hat{p}_i^{*(B)}$.

The simplest $100(1 - \alpha)$percent bootstrap confidence interval approximation for p_i is given by

$$\left(\hat{p}_i^{*(B\alpha/2)}, \hat{p}_i^{*(B(1-\alpha)/2)} \right).$$

Rosenberger and Hu show that this simple confidence interval provides near perfect coverage. The same techniques can be used for measures of treatment differences and can incorporate delayed response and staggered entry in the simulation phase 2.

11.3 POWER

Response-adaptive randomization induces additional correlation among the responses, and this leads to an increase in the variance of the test statistic. This increased variance contributes to a decrease in power for standard tests based on a population model. In this section, we explore the power of response-adaptive randomization procedures.

11.3.1 The relationship between power and the variability of the design

In general, for clinical trials of two treatments, power of the test will be intimately linked to $\text{Var}(N_A(n)/n)$. One can see this quite readily when examining the noncentrality parameter for the test of the simple difference in binary response trials (Hu and Rosenberger, 2003).

Suppose we have a fixed target proportion on treatment A, ρ; for instance, ρ could be based on some optimization criterion, or the limiting allocation of an urn design, as discussed in Chapter 10. For this case, we can calculate the noncentrality parameter for the Z-test as follows:

$$\frac{(p_A - p_B)^2}{p_A q_A / N_A(n) + p_B q_B / N_B(n)},$$

which can be rewritten as

$$\frac{(p_A - p_B)^2}{p_A q_A / [n\rho + n(N_A(n)/n - \rho)] + p_B q_B / [n(1 - \rho) - n(N_A(n)/n - \rho)]}.$$

Now define a function

$$g(x) = \frac{(p_A - p_B)^2}{p_A q_A / (\rho + x) + p_B q_B / ((1 - \rho) - x)}.$$

We have the following expansion:

$$g(x) = g(0) + g'(0)x + g''(0)x^2/2 + o(x^2).$$

After some calculation, we have that the noncentrality parameter of the test is given by

$$n \times \left\{ \frac{(p_A - p_B)^2}{p_A q_A / \rho + p_B q_B / (1 - \rho)} \right.$$

$$+ (p_A - p_B)^2 \frac{(p_A q_A (1 - \rho)^2 - p_B q_B \rho^2)}{(p_A q_A (1 - \rho) + p_B q_B \rho)^2} (N_A(n)/n - \rho)$$

$$- (p_A - p_B)^2 \frac{p_A q_A p_B q_B}{((1 - \rho)\rho)^3} (N_A(n)/n - \rho)^2$$

$$\left. + o((N_A(n)/n - \rho)^2) \right\}. \tag{11.13}$$

The first term in (11.13) is determined by ρ and represents the noncentrality parameter for the fixed design. The second term in (11.13) represents the bias of the randomized design from the target proportion. With the design shifting to a different side from the target proportion ρ, the noncentrality parameter will increase or decrease according the coefficient

$$(p_A - p_B)^2 \frac{(p_A q_A (1 - \rho)^2 - p_B q_B \rho^2)}{(p_A q_A (1 - \rho) + p_B q_B \rho)^2}.$$

To control the power, it may be desired to have this coefficient be 0. It is interesting to see that this coefficient equals 0 if and only if $p_A q_A (1 - \rho)^2 - p_B q_B \rho^2 = 0$, that is,

$$\rho = \frac{\sqrt{p_A q_A}}{\sqrt{p_A q_A} + \sqrt{p_B q_B}},$$

that is, Neyman allocation.

For response-adaptive randomization procedures, we can consider the expectation of the noncentrality parameter. If $E(N_A(n)/n - \rho) = 0$, at least to order $o(1/n)$, then the average power lost using the response-adaptive randomization procedure is then a function of

$$-(p_A - p_B)^2 \frac{p_A q_A p_B q_B}{((1 - \rho)\rho)^3} E(N_A(n)/n - \rho)^2,$$

which represents the variability of the design.

Thus, we can use the variance of $N_A(n)/n$ to compare the power of response-adaptive randomization procedures with same allocation limit or the variance and bias if they do not have the same limiting allocation. Hu and Rosenberger (2006) demonstrate how to derive the asymptotic variance of $N_A(n)/n$ for both sequential estimation-type procedures and urn models and give examples of how to compute it for 13 response-adaptive randomization procedures.

11.3.2 Asymptotically best procedures

Hu, Rosenberger, and Zhang (2006) determine a Rao–Cramér lower bound on the variance of $N_A(n)/n$ as follows. If there exists a positive definite matrix $V(\theta)$ such that

$$\sqrt{n}\left(\frac{N(n)}{n} - \rho(\theta)\right) \to N(\mathbf{0}, V(\theta))$$

in distribution, then the Rao–Cramér lower bound on $V(\theta)$ is given by

$$\frac{\partial \rho(\theta)}{\partial \theta} I^{-1}(\theta) \frac{\partial \rho(\theta)'}{\partial \theta}. \tag{11.14}$$

Using this lower bound, Hu, Rosenberger, and Zhang (2006) call a response-adaptive randomization procedure *asymptotically best* if, for a desired target allocation $\rho(\theta)$, $V(\theta)$ attains the lower bound. The discovery of a fully randomized procedure that is asymptotically best for any target allocation is referred to as the *fundamental question of response-adaptive randomization* by Hu and Rosenberger (2006).

As most urn models can only target a single limiting allocation ($\rho = (q_B/(q_A + q_B), q_A/(q_A + q_B))$), the Rao–Cramér bound for procedures targeting ρ is computed from (11.14) as

$$\frac{q_A q_B (p_A + p_B)}{(q_A + q_B)^3}. \tag{11.15}$$

Ivanova (2003) shows that the drop-the-loser rule attains this bound and is always less variable than the randomized play-the-winner rule. Thus, the drop-the-loser rule is the asymptotically best procedure for targeting $\rho = (q_B/(q_A + q_B), q_A/(q_A + q_B))$. However, this target allocation is not optimal for preserving power. One might instead wish to target optimal allocations given in Table 10.1. For instance, the lower bound for the target

$$\rho = \left(\frac{\sqrt{p_A}}{(\sqrt{p_A} + \sqrt{p_B})}, \frac{\sqrt{p_B}}{(\sqrt{p_A} + \sqrt{p_B})}\right)$$

can be computed as

$$\frac{1}{(\sqrt{p_A} + \sqrt{p_B})^3}\left(\frac{p_B q_A}{\sqrt{p_A}} + \frac{p_A q_B}{\sqrt{p_B}}\right).$$

The doubly adaptive biased coin design does not attain this bound.

11.3.3 Response-adaptive randomization and sequential monitoring

The role and importance of sequential monitoring in clinical trials are described in 6.14. In the population model-based context, sequential monitoring requires the distribution of sequentially computed test statistics, a stopping boundary (typically determined using a spending function), and the computation of the Fisher's information. The goal is to preserve the type I and type II error rates at the end of the trial. Because of the complex dependence induced by response-adaptive randomization, computing the joint distribution of sequential tests is challenging.

For the randomized play-the-winner rule, Stallard and Rosenberger (2002) computed the exact distribution of sequentially computed tests by using a networking algorithm. They found that the randomized play-the-winner rule does not improve expected treatment successes when used in conjunction with a particular sequential monitoring scheme.

For the doubly adaptive biased coin design, Zhu and Hu (2010) found the joint asymptotic distribution of sequentially computed test statistics. They demonstrate by simulation, for different target allocations and different spending functions, that the size and power of the sequential testing procedure is preserved similarly to complete randomization. However, the expected total number of failures was less using response-adaptive randomization. They also show that the doubly adaptive biased coin design has a simple form of information: the information fraction is the proportion of the sample size enrolled; they conjecture that this form is true for most response-adaptive randomization procedures.

11.4 RANDOMIZATION-BASED INFERENCE

In principle, randomization-based inference can be conducted following response-adaptive randomization, as with any restricted randomization procedure, as described in Chapter 6. The responses are treated as deterministic, and all possible permutations of the treatment assignment sequence, for the given set of responses, are enumerated, along with their associated probabilities. Under the null hypothesis of no treatment effect, responses should be evenly distributed across all possible treatment sequences; deviations are evidence of a treatment effect.

It should be noted that this approach is not consistent with the theory of permutation tests (*e.g.*, Pesarin (2001)). The validity of permutation tests (which rely on resampling, but not re-randomization) is based on conditioning on the data as a sufficient statistic. In restricted randomization, the responses, $Y_1, \dots, Y_n$, form a sufficient statistic, but in response-adaptive randomization, the responses *and* treatment assignments are jointly sufficient; that is, $T_1, \dots, T_n, Y_1, \dots, Y_n$. Conditioning on the entire set of sufficient statistics does not make sense in this context.

Nevertheless, many authors have proposed using randomization-based inference following response-adaptive randomization (cf. Wei, 1988; Rosenberger, 1993; Simon and Simon, 2011; Galbete and Rosenberger, 2015). The distribution of the randomization test is based on the joint distribution of $T_1, \dots, T_n$, conditioned on observed values $y_1, \dots, y_n$, unlike in restricted randomization where the treatment assignment distribution is independent of the responses. The Monte Carlo re-randomization test procedure is identical to that described in Section 6.9.

As with restricted randomization procedures, randomization-based inference can be performed following a response-adaptive randomization procedure using the family of linear rank tests. When $K = 2$, the linear rank test is given by

$$W = \sum_{j=1}^{n} (a_{jn} - \bar{a}_n)T_j, \tag{11.16}$$

where the a_{jn}'s are fixed constants, $T_j = 1$ if treatment A was assigned, and 0 if treatment B. In the case of binary response, we can let a_{jn} be 1 if success and 0 if failure. In this case, letting $S_A(n)$ and $S_B(n)$ be the number of successes on A and B, respectively, a little algebra shows that (11.16) is equivalent to

$$S = \frac{N_A(n)N_B(n)}{n} \left(\frac{S_A(n)}{N_A(n)} - \frac{S_B(n)}{N_B(n)} \right). \tag{11.17}$$

Note that this test depends on the jointly sufficient statistics $(S_A(n), S_B(n), N_A(n))$.

Table 11.1 shows the computation of the exact test under RPW(1, 1) randomization for $n = 4$ when the patient's responses were $a_{jn} = \{1, 0, 0, 1\}$ and the observed allocation was $T_j = \{A, B, B, A\}$. Then the observed test statistic is $S_{obs.} = 1.0$. The unconditional p-value is computed by summing the probabilities of each sequence where $|S_l| \geq 1.0$. This yields $p_u = 2/30$.

Table 11.1 *Unconditional reference set for computation of the linear rank test following* RPW(1, 1) *randomization.*

Sequence (l)	$\Pr(L = l)$	S_l
AAAA	1/15	0.0
AAAB	1/10	−0.5
AABA	1/10	0.5
AABB	1/15	0.0
ABAA	3/40	0.5
ABAB	1/20	0.0
ABBA	1/30	1.0
ABBB	1/120	0.5
BAAA	1/120	−0.5
BAAB	1/30	−1.0
BABA	1/20	0.0
BABB	3/40	−0.5
BBAA	1/15	0.0
BBAB	1/10	−0.5
BBBA	1/10	0.5
BBBB	1/15	0.0

Wei (1988) proposed a version of this test with uncentered scores, given by

$$S = \sum_{j=1}^{n} a_{jn} T_j = S_A(n) \tag{11.18}$$

and gives an algorithm to compute the exact p-value. However, this is not a function of the joint sufficient statistics, so it does not contain sufficient information to serve as a test of the treatment effect. This test generated much controversy, which was recorded in the paper by Begg (1990) with ensuing discussion. The test in (11.17) certainly lends itself to a more straightforward interpretation in terms of the observed treatment difference and uses the entire set of joint sufficient statistics.

11.5   PROBLEMS

11.1   Use the delta method to show (11.12).

11.2   Derive (11.13).

11.3   Use (11.9) and (11.8) to show that the Rao–Cramér lower bound for a response-adaptive randomization procedure targeting

$$\rho = \left(\frac{q_B}{q_A + q_B}, \frac{q_A}{q_A + q_B} \right)$$

is given by (11.15).

11.4   Read Wei (1988) and Begg (1990), along with the ensuing discussion. Write a short paper summarizing the various methods discussed and points for and against each (from Begg and the discussants). What are your views on inference following a response-adaptive randomization procedure?

11.6   REFERENCES

BAI, Z. D., HU, F., AND ROSENBERGER, W. F. (2002). Asymptotic properties of adaptive designs with delayed response. *Annals of Statistics* **30** 122–139.

BEGG, C. B. (1990). On inferences from Wei's biased coin design for clinical trials. *Biometrika* **77** 467–484 (with discussion).

COAD, D. S. AND GOVINDARAJULU, Z. (2000). Corrected confidence intervals following a sequential adaptive trial with binary response. *Journal of Statistical Planning and Inference* **91** 53–64.

COAD, D. S. AND IVANOVA, A. (2001). Bias calculations for adaptive urn designs. *Sequential Analysis* **20** 229–239.

COAD, D. S. AND WOODROOFE, M. B. (1997). Approximate confidence intervals after a sequential clinical trial comparing two exponential survival curves with censoring. *Journal of Statistical Planning and Inference* **63** 79–96.

GALBETE, A. AND ROSENBERGER, W. F. (2015). On the use of randomization tests following adaptive designs. *Journal of Biopharmaceutical Statistics*, in press.

HU, F. AND ROSENBERGER, W. F. (2003). Optimality, variability, power: evaluating response-adaptive randomization procedures for treatment comparisons. *Journal of the American Statistical Association* **98** 671–678.

HU, F. AND ROSENBERGER, W. F. (2006). *The Theory of Response-Adaptive Randomization in Clinical Trials*. John Wiley & Sons, Inc., New York.

HU, F., ROSENBERGER, W. F., AND ZHANG, L.-X. (2006). Asymptotically best response-adaptive randomization procedures. *Journal of Statistical Planning and Inference* **136** 1911–1922.

HU, F. AND ZHANG, L.-X. (2004). Asymptotic normality of adaptive designs with delayed response. *Bernoulli* **10** 447–463.

IVANOVA, A. V. (2003). A play-the-winner type urn model with reduced variability. *Metrika* **58** 1–13.

LEHMANN, E. L. (1983). *The Theory of Point Estimation*. John Wiley & Sons, Inc., New York.

MATTHEWS, P. C. AND ROSENBERGER, W. F. (1997). Variance in randomized play-the-winner clinical trials. *Statistics and Probability Letters* **35** 223–240.

MAY, C. AND FLOURNOY, N. (2009). Asymptotics in response-adaptive designs generated by a two-color, randomly reinforced urn. *Annals of Statistics* **37** 1058–1078.

PESARIN, F. (2001). *Multivariate Permutation Tests: With Applications in Biostatistics*. John Wiley & Sons, Inc., New York.

ROSENBERGER, W. F. (1993). Asymptotic inference with response-adaptive treatment allocation designs. *Annals of Statistics* **21** 2098–2107.

ROSENBERGER, W. F., FLOURNOY, N., AND DURHAM, S. D. (1997). Asymptotic normality of maximum likelihood estimators from multiparameter response-driven designs. *Journal of Statistical Planning and Inference* **60** 69–76.

ROSENBERGER, W. F. AND HU, F. (1999). Bootstrap methods for adaptive designs. *Statistics in Medicine* **18** 1757–1767.

SIMON, R. AND SIMON, N. R. (2011). Using randomization tests to preserve type I error with response adaptive and covariate adaptive randomization. *Statistics and Probability Letters* **81** 767–772.

STALLARD, N. AND ROSENBERGER, W. F. (2002). Exact group-sequential designs for clinical trials with randomised play-the-winner allocation. *Statistics in Medicine* **21** 467–480.

WEI, L. J. (1988). Exact two-sample permutation tests based on the randomized play-the-winner rule. *Biometrika* **75** 603–606.

WEI, L. J., SMYTHE, R. T., LIN, D. Y., AND PARK, T. S. (1990). Statistical inference with data-dependent treatment allocation rules. *Journal of the American Statistical Association* **85** 156–162.

ZHU, H. AND HU, F. (2010). Sequential monitoring of response-adaptive randomized clinical trials. *Annals of Statistics* **38** 2218–2241.

12

Response-Adaptive
Randomization in Practice

12.1 BASIC ASSUMPTIONS

In this chapter, we explore practical considerations in the use of response-adaptive randomization. It should be clear from Chapters 10 and 11 that there are three basic assumptions underlying the use of these types of designs.

First, one must assume that it is feasible to identify the "better" treatment with high probability. This, in turn, will depend on the target sample size for the trial and the treatment effect anticipated. Usually, the anticipated treatment effect is modest, since studies are designed to detect the minimal clinically relevant difference in treatments. The smaller the designed effect, the larger the sample size needed to provide a high probability that the better of the two treatments is so identified.

Second, we must assume that the "better" treatment is not associated with any potential severe toxicity, short- or long-term. Otherwise, the design will be assigning the majority of patients to an unsafe therapy. In fact, some have suggested the importance of at least beginning the trial with equal allocation until some experience is gained that the treatments are safe, before beginning adaptive randomization.

Third, some patient data on the primary outcome of the trial must be accrued prior to randomizing most of the patients. This immediately eliminates long-term survival trials with limited recruitment and a follow-up extending years. In many of those trials, outcome data become evident only after the recruitment phase has ended. While long-term survival trials represent a large portion of major multicenter clinical trials, there are many shorter duration trials in which the recruitment period can be extended to provide data for the adaptation of future allocation probabilities.

These criteria, particularly the second, would tend to preclude use of response-adaptive randomization in most phase II trials of new drugs where the safety of the agent has yet to be established. However, response-adaptive randomization

Randomization in Clinical Trials: Theory and Practice, Second Edition.
William F. Rosenberger and John M. Lachin.
© 2016 John Wiley & Sons, Inc. Published 2016 by John Wiley & Sons, Inc.

could be ideal in phase II studies of an established safe agent in a new patient population. Likewise, this would preclude the use of response-adaptive randomization in phase III trials of agents in which animal toxicology or phase II studies have raised the possibility of short- or long-term adverse effects. However, response-adaptive randomization could be ideal for studies of competing agents for a given indication, all of which were previously documented to be safe, or for a phase III compound that is a member of a family of drugs, the safety of which has already been established.

While most of the models examined in Chapter 10 assume that patient responses are ascertainable immediately before the next patient is randomized, that assumption is used only to simplify the probabilistic properties of the response-adaptive randomization rules. In practice, one can "adapt" at certain fixed points in the trial using grouped data already accrued, or one can factor in a delayed response by just using the data available. In the latter setting, one would update the urn (for urn models) or update the maximum likelihood estimators (for sequential maximum likelihood procedures) as each patient responds. Simulation studies have shown that (at least for urn models), whereas the allocation probabilities are not as extreme as for immediate response trials, response-adaptive randomization with delayed response still reduces the expected number of failures and puts more patients on the better treatment when there is delayed response (Rosenberger, 1999).

While certainly a minority of clinical trials are performed with a primary outcome that is ascertainable immediately, a good number of such trials are conducted. Often these are clinical trials of surgical interventions or other medical procedures with an easily ascertainable "success" or "failure" outcome, which is known before the next patient undergoes the procedure. One example is the prevention of hypotension associated with spinal anesthesia for Cesarean section. Rout, *et al.* (1993) describe such a trial of crystalloid preload versus placebo, using Zelen's play-the-winner rule (Section 10.2.1) to allocate treatments.

12.2 BIAS, MASKING, AND CONSENT

Because response-adaptive randomization procedures are randomized, they enjoy many of the same benefits of other randomization procedures in terms of mitigation of bias. However, there are several ways that bias can enter a trial using response-adaptive randomization procedures.

As with any randomization procedure, the clinical trial should, whenever possible, be double-masked. The current allocation probabilities should be kept strictly confidential by the statistician responsible for randomization, as knowledge of the allocation probabilities is tantamount to knowledge of the current treatment effect. Even in unmasked studies, response-adaptive randomization procedures offer some protection from selection bias provided that the responses of previously entered patients are masked. If the responses are unmasked, and their corresponding treatment assignments also unmasked, then one would expect that such designs afford less protection against selection bias than restricted randomization procedures, if the shift in probability of assignment to A away from 0.5 is larger for response-adaptive randomization when one treatment is superior to that of restricted randomization when there is a treatment imbalance.

These procedures also provide some protection against accidental bias, provided that one assumes that all subjects arise at random from an underlying homogeneous population, such that the probability of the covariates, and also patient responses, are identical over time. For example, consider the following simple scenario. Assume that a simple two-stage adaptive procedure is employed with the same number of subjects recruited in the first and second stages. In the first stage, the probability of assignment to A is 0.5 and the probability of recruiting a male subject is 0.5. Then at the second stage, based on the finding of more beneficial response with A during the first stage, the probability of assignment to A is modified to 0.8. Now also assume that by chance or due to a change in recruitment strategies, the probability of recruiting a male subject during the second stage is 0.7. Thus, during the second stage, it is more likely that a patient will be male than female and more likely that the patient will be assigned to A rather than B. This will result in a covariate imbalance, in which 62.3 percent of those assigned to A will be male versus 55.7 percent of those assigned to B. If the probability of treatment response differed greatly between males and females, this would also introduce a bias into the results of the study. One could also evaluate by simulation the susceptibility of response-adaptive randomization to the trend of a covariate, qualitative or quantitative, over time. However, this simple example, with an extreme shift in the assignment probabilities, and an extreme shift in the covariate distribution, still results in a degree of imbalance that would be readily adjusted for in a *post hoc* stratification or a regression adjustment, as described in Chapter 7.

Rosenberger and Lachin (1993) suggest that consent forms should state that participants will receive one of two treatments and the probability of treatment assignments will depend on the relative merits of the two treatments based on responses of previously treated volunteers. Such a statement should make the trial more attractive to participants than simply telling them that they are equally likely to receive either treatment. It should also be made clear that the treatment performing better thus far may not, in fact, be the better treatment overall, because the study has not been completed and there are not enough patients currently to make that evaluation.

Informing the patient of the nature of the response-adaptive randomization in this way may lead to a different kind of bias, coined *accrual bias* by Rosenberger (1996), in which volunteers may wish to be recruited later in the trial so as to benefit from the full impact of previous outcomes, and thereby have a better chance to receive the better treatment. Rosenberger (1999) recommends that patients be masked to their sequence number in the trial to prevent accrual bias; whether such masking is acceptable to patients and physicians has not been investigated. Accrual bias is irrelevant in trials dealing with emergency therapies, such as emergency surgical procedures.

12.3 LOGISTICAL ISSUES

There are two main differences between the implementation of response-adaptive randomization and the implementation of other randomization procedures. These differences become magnified as the complexity of the trial increases, particularly in the multicenter situation. First, as pointed out by Faries, Tamura, and Andersen (1995), response-adaptive randomization requires much more communication among the sponsor or coordinating center and the investigators in a multicenter clinical trial. In particular, the randomization procedure must be updated as each patient response is received. Faries, Tamura, and Andersen found that some investigators did not always call in response data after a patient was randomized, and clinical trials personnel had to prompt investigators for missing data in order to update the randomization. Secondly, since the randomization must be dynamically updated, it is not possible to generate the randomization sequence in advance. The investigator cannot assign packaged drug sequentially and must contact the coordinating center or sponsor for the proper packages (if they are prepackaged) for each individual patient. As discussed in Section 8.6.1, since we do not know exactly the number of patients to be assigned to each treatment, there will need to be some oversupply in packaging. Faries, Tamura, and Andersen found that the system worked reasonably well in adaptive clinical trials they ran, but it required additional resources. They had two research associates on-call at all times since some investigators randomized patients on weekends and after hours and called in to get the randomization assignment.

Stratification is straightforward with response-adaptive randomization, as it is with restricted randomization procedures. One simply produces a separate randomization sequence within each of the strata. In particular, for urn models, one can run a separate urn within each stratum.

Faries, Tamura, and Andersen (1995, p. 5) concluded that

> ... We feel that the only way to gain experience [with response-adaptive randomization] is to conduct such trials and learn from our successes and failures. We encourage our clinical colleagues in the biopharmaceutical industry to do the same.

12.4   SELECTION OF A PROCEDURE

Choosing to implement a response-adaptive randomization procedure will require additional time and effort from the statistician both to select an appropriate randomization procedure and to fuel understanding by scientific colleagues in the clinical trial. In many clinical trials where there is a rush to determine an appropriate protocol, the effort required cannot be reasonably accomplished. Selection of a procedure requires simulation of the procedure under various possible clinical trial conditions. There are several aspects that the statistician should investigate:

1. Under a realistic model of the patient responses, will the response-adaptive randomization procedure work as intended? Will more patients, on average, be assigned to the superior treatment? Is the variability of the procedure within reasonable limits?
2. What is the required sample size to maintain a reasonable level of power for the study? If this sample size is larger than that required for equal allocation, are there

really any savings in terms of expected numbers of treatment failures or expected numbers of patients assigned to the inferior treatment?

3. What if there is a drift in patient characteristics over time? Will this affect the adaptation adversely or introduce a covariate imbalance?

When simulating sample size and power, we have found that the easiest way is to compute the sample size n^* required for a standard clinical trial with equal allocation under an alternative reflective of the clinically relevant treatment effect , as discussed in Section 2.6. Then the response-adaptive randomization procedure is simulated k times with n^* patients, and the proportion of the k times the test statistic rejects the null hypothesis is then the simulated average power of the procedure. If the procedure is less powerful than a restricted randomization procedure, one then increases n^* and reruns the simulation until the power is similar. One then also simulates the expected number of treatment failures or the expected number assigned to the inferior treatment and compares this value to that obtained with restricted randomization. This is the approach used in Hu and Rosenberger (2006, Chapter 8).

By using sophisticated data structures, such as priority queues, one could also incorporate delayed response into the simulation, by assuming that arrivals are staggered, perhaps according to a uniform distribution, as discussed in Section 2.5, and response is delayed according to some time-to-event distribution. Patient entries and responses are then followed through a queuing system; this can be programmed using a priority queue (see, e.g., Rosenberger and Seshaiyer 1997; Rosenberger and Hu, 1999).

Flournoy, Haines, and Rosenberger (2013) describe how to use trade-off plots, in the same spirit as in Section 8.4, to select an appropriate response-adaptive randomization procedure. They plot the efficiency of estimation versus an ethical criterion. They show that each criterion is a function of the sufficient statistics. In the binary case, this is $(S_A(n), S_B(n), N_A(n))$ (see Section 11.2.2). One can simulate the sufficient statistics under a particular model for (p_A, p_B), and then one can compute the criteria as functions of the sufficient statistics. For example, one measure of efficiency is the mean squared error of the simple difference metric, given by

$$\left[E \left(\left(\frac{S_A(n)}{N_A(n)} - \frac{S_B(n)}{n - N_A(n)} \right) - (p_A - p_B) \right) \right]^2 + \mathrm{Var} \left(\frac{S_A(n)}{N_A(n)} - \frac{S_B(n)}{n - N_A(n)} \right).$$

Measures of ethical consequence may be the expected proportion of patients assigned to the inferior treatment, $E(N_B(n)/n)$, (if B is inferior), or the total failure rate $E((n - S_A(n) - S_B(n))/n)$. Unfortunately, these metrics cannot be scaled similarly. However, one can compare procedures directly for different values of (p_A, p_B) and see which procedure is closest to the origin. They find that, while the parameterization of the various response-adaptive randomization procedures does not make much difference, procedures that target an optimal allocation tend to perform better than those that do not. They also examine several procedures for normally distributed responses.

For binary response trials with fairly immediate response, the doubly adaptive biased coin design with allocation function by Hu and Zhang (2004), or the ERADE, discussed in Sections 10.4.2 and 10.4.4, targeting an appropriate R^* in Table 10.1 appears to be the most powerful procedure with the maximum savings of patients, when p_A and p_B are less than 0.5. If treatments are believed to be more successful than 0.5, then the drop-the-loser rule in Section 10.5.4 has proven to be the most powerful. The randomized play-the-winner rule is particularly variable when $p_A + p_B > 3/2$ and resultant losses in power make it unattractive. For continuous and survival outcomes, the methodology described in Zhang and Rosenberger (2006, 2007) preserves power at a modest benefit to patients. The techniques of Tymofyeyev, Rosenberger, and Hu (2007) can be used for $K > 2$ treatments and binary responses.

12.5 BENEFITS OF RESPONSE-ADAPTIVE RANDOMIZATION

The potential benefits of adaptive allocation for clinical trials were recognized quite early. In 1969, Cornfield, Halperin, and Greenhouse (1969, p. 760) wrote:

> Application of these results might ease the ethical problem involved in trials on human subjects. The usual ethical justification for not administering an agent of possible efficacy to all patients is the absence of definite information about its effectiveness. However satisfactory this justification may be before the trial starts it rapidly loses cogency as evidence for or against the agent accumulates during the course of the trial. But any solution … which permits adaptive behavior … at least reduces this ethical problem.

Weinstein's (1974) special article in *New England Journal of Medicine* strongly advocated adaptive allocation as an alternative to traditional treatment assignment rubrics (pp. 1279, 1284):

> … Any decision rule for allocating patients to clinical procedures in any way other than according to the best interest of the subject at hand does entail a sacrifice on the part of the subject …. Adaptive methods should be used as a matter of course. It never pays to commit oneself to a protocol under which information available before the study or obtained during its course is ignored in the treatment of a patient.

Byar *et al.* (1976) responded to Weinstein's article by pointing out many of the subtle problems with adaptive designs. The comments are extremely cogent, especially in light of the limited existing literature on the subject at the time. They point out the potential for biases with time heterogeneity, the potential loss of power due to unequal sample sizes, and the difficulty of applying the methodology to long-term trials of chronic diseases.

Other authors have argued heatedly against any form of response-adaptive *randomization*. Royall (1991, p. 58) writes:

> … The ethical problems are clear: after finding enough evidence favoring [treatment] A to require reducing the probability of [treatment] B, the physician … must see that the next patient gets A, not just with high probability, but with certainty.

This point was argued extensively in discussion to Royall's paper; see particularly the response of Byar, Simon (1991) writes:

> [I do] not find it attractive to approach a patient saying that I do not know which treatment is better, but treatment *A* is doing better therefore I will give you a greater than 50 percent chance of getting it.

While response-adaptive randomization does not eliminate the ethical problem of randomizing patients to the inferior treatment, it mitigates it by making the probability of assignment to the inferior treatment smaller. We find this to be an attractive alternative to the usual 50 : 50 randomization procedures for certain clinical trials. We believe patients and physicians will find it attractive too. Many clinical trials have used balanced allocation to multiple treatment arms as a successful recruitment tool; for example, a trial of three experimental therapies versus a placebo or a trial with a combination therapy arm, two single-therapy arms, and a placebo. In such cases, one can advertise that patients have a 75 percent chance of being assigned to an experimental arm. Response-adaptive randomization can be used similarly as a recruitment tool. In truth, patients do not enter clinical trials in order to be on a placebo (although many patients would prefer to be assigned to the placebo in cases where there may be some risk of adverse events).

While Royall's point advocates deterministic assignments based on an adaptive procedure, such studies are prone to the biases of nonrandomized studies. We prefer to maintain the benefits of randomization while increasing the number of patients assigned to the superior treatment, if it exists. Tamura, *et al.*, (1994, p. 775) give the following reason for the controversy around response-adaptive randomization:

> We believe that because [response-adaptive randomization] represents a middle ground between the community benefit and the individual patient benefit, it is subject to attack from either side.

Following an adaptive clinical trial on fluoxetine for depression, (see Section 12.6.2), the investigators reported (Tamura, *et al.*, (1994, p. 775)):

> We were encouraged by the cooperation and willingness of our clinical research colleagues and our investigators to design, implement, and report on such a trial.... This has encouraged us to continue research efforts into both the implementation and analysis of adaptive trials.

12.6   SOME EXAMPLES

12.6.1   The extracorporeal membrane oxygenation trial

The randomized play-the-winner rule was used in a clinical trial of extracorporeal membrane oxygenation (ECMO; Bartlett, *et al.*, 1985), a surgical procedure for newborns with respiratory failure. The technique had been used when infants were moribund and unresponsive to conventional treatment (ventilation and pharmaco-logic therapy). Early trials on safety and efficacy had indicated that the ECMO technique was safe and had an overall success rate of 56 percent, compared to a success rate of about 20 percent for conventional therapy. Bartlett, *et al.* (1985, p. 480) state that the $RPW(1, 1)$ rule was chosen for the following reasons:

> (1) the outcome of each case [was] known soon after randomization, making it possible to use; (2) [it was] anticipated that most ECMO patients would survive and most control patients would die, so significance could be reached with a modest number of patients; [and] (3) it was a reasonable approach to the sci-entific/ethical dilemma.

In the randomization scheme, the first patient was assigned to ECMO and sur-vived, changing the urn composition to 2 ECMO balls and 1 control ball. The second patient was assigned to conventional therapy and died, leading to 3 ECMO balls and 1 control ball. Each subsequent randomization was to ECMO, and each of the patients survived. The trial was stopped after 12 total patients, using a stopping rule described by Cornell, Landenberger, and Bartlett (1986).

Serious questions arose about the validity of such a trial. The foremost question raised is whether two treatments can adequately be compared when only one patient was assigned to one of the treatments. The validity of clinical trials with a sample size of 12 has also been questioned. In any event, the clinical trials were not convincing and led to at least two other clinical trials of the same therapy (O'Rourke, *et al.* (1989); see also Ware (1989); UK Collaborative ECMO Trial Group (1996)).

What went wrong? We know from Chapter 10 that the $RPW(1, 1)$ rule is highly variable, particularly when $p_A + p_B > 3/2$, when the variance depends on the initial composition of the urn. In retrospect, starting with more than one ball of each type should have resulted in more patients on the control arm, and a minimum sample size should have been set in advance. To this day, some investigators use the ECMO example as a reason not to perform response-adaptive randomization at all. This is unfortunate because we think this is exactly the type of trial for which response-adaptive randomization would be particularly advantageous.

12.6.2 The fluoxetine trial

The RPW(1, 1) rule was employed in a clinical trial of fluoxetine versus placebo for depressive disorder. The trial was stratified by normal and shortened rapid eye move-ment latency (REML), so two urns were used in the randomization. In order to avoid an ECMO-like situation with too few controls, the first six patients in each stratum were assigned using a permuted block design. The primary outcome, a reduction of 50 percent or greater on the Hamilton Depression Scale between baseline and final active visit after a minimum of 3 weeks of therapy could only be ascertained after approximately 8 weeks. Determining that this was too long a period in which to run an adaptive trial, investigators used a surrogate outcome to update the urn. They defined a surrogate responder as a patient exhibiting a reduction greater than 50 percent on the

Hamilton Depression Scale in two consecutive visits after at least 3 weeks of therapy. The trial was stopped after 61 patients had responded in accordance with the surrogate criterion; the trial randomized a total of 89 patients: 21 fluoxetine patients and 20 placebo patients in the shortened REML stratum; 21 fluoxetine and 21 placebo patients in the normal REML stratum. Six patients did not have a final outcome status. A significant treatment effect was found in the shortened REML category, but not the normal REML stratum. The primary outcome was analyzed using a Monte Carlo randomization-based analysis. Although there was a significant treatment effect in the shortened REML stratum, the randomization did not favor the treatment arm. The investigators found that the randomization sequence tended to assign patients to placebo when the probability of allocation to fluoxetine was higher. They found that the probability of their particular sequence, given the allocation probabilities, was about 22 percent. (See Tamura, *et al.*, 1994.)

12.7 CONCLUSIONS

Response-adaptive randomization procedures require more work to implement, in that the randomization procedure must be programmed and the program must update the allocation probabilities after each patient response. They also require much work on the part of the statistician in the design phase of the trial. We recommend that extensive simulations be run to ascertain the operating characteristics of the procedure, to determine sample size requirements, and to assess the potential benefits of using response-adaptive randomization. The fluoxetine trial is an example of a well-conducted and thoughtfully designed clinical trial. However, the added benefit to patients was minimal, because the allocation was close to equal even in the stratum where there was a treatment effect.

Rosenberger (1999) discusses conditions under which the use of response-adaptive randomization is reasonable. We note some of them here:

1. The therapies have been evaluated previously for toxicity. This is important to ensure that the response-adaptive randomization does not place more patients on a highly toxic treatment.
2. Delay in response is moderate, allowing the adaptation to take place.
3. Duration of the trial is limited and recruitment can take place during most or all of the trial.
4. The trial is carefully planned with extensive simulations run under different models.
5. The experimental therapy is expected to have significant benefits to the public health.
6. Modest gains in terms of treatment successes are desirable from an ethical standpoint.

Few areas of statistics have contributed to more controversy than response-adaptive randomization (see Problem 12.1). However, the extra effort required to design and implement clinical trials using response-adaptive randomization could result in significant benefits to patients and clinical medicine in general.

12.8 PROBLEMS

12.1 a. Familiarize yourself with the two famous ECMO trials by looking at the original papers (Bartlett, *et al.*, 1985; O'Rourke, *et al.*, 1989).
b. Now read about the controversy that ensued in the following papers and attendant discussions (Ware, 1989; Royall, 1991).
c. Now familiarize yourself with the 1996 UK Collaborative ECMO Trial and read the accompanying editorial (UK Collaborative ECMO Trial Group, 1996).
d. Write a 15 minute position paper to be presented in a class debate on the three ECMO trials. Focus on the following issues:

(i) Were the three trials necessary? If not, what were the appropriate alternatives?
(ii) Should a response-adaptive design have been used for this type of trial? If so, which one and why?
(iii) Should response-adaptive randomization ever be used? Under what conditions?
(iv) Is randomization necessary? Are clinical trials ethical? Focus in particular on the interchange between Royall and Byar in the Royall (1991) paper.
(v) Was the 1995 UK Collaborative Trial ethical?

12.2 Find a clinical trial in a major medical journal (*e.g.*, *New England Journal of Medicine, Journal of the American Medical Association, Lancet*) for which response-adaptive randomization would be appropriate. Write a short paper explaining why this would be an appropriate trial and describing procedures and statistical considerations in redesigning the trial using response-adaptive randomization.

12.3 For the randomization procedure in Problem 12.2, find, by simulation, the sample size necessary to attain 90 percent power for a specific alternative of interest.

12.4 For the scenario described in Section 12.2, wherein the probability of assignment to treatment A shifts from 0.5 to 0.8 in the first and second stages of recruitment, and the probability of recruiting a male shifts from 0.5 to 0.7, for equal numbers recruited in both stages show that the probability of a male is 0.623 in treatment group A and 0.557 in treatment group B.

12.9 REFERENCES

Bartlett, R. H., Roloff, D. W., Cornell, R. G., Andrews, A. F., Dillon, P. W., and Zwischenberger, J. B. (1985). Extracorporeal circulation in neonatal respiratory failure: a prospective randomized study. *Pediatrics* **76** 479–487.

BYAR, D. P., SIMON, R. M., FRIEDEWALD, W. T., SCHLESSELMAN, J. J., DEMETS, D. L., ELLENBERG, J. H., GAIL, M. H., AND WARE, J. H. (1976). Randomized clinical trials: perspectives on some recent ideas. *New England Journal of Medicine* **295** 74–80.

CORNELL, R. G., LANDENBERGER, B. D., AND BARTLETT, R. H. (1986). Randomized play the winner clinical trials. *Communications in Statistics—Theory and Methods* **15** 159–178.

CORNFIELD, J., HALPERIN, M., AND GREENHOUSE, S. W. (1969). An adaptive procedure for sequential clinical trials. *Journal of the American Statistical Association* **64** 759–770.

FARIES, D. E., TAMURA, R. N., AND ANDERSEN, J. S. (1995). Adaptive designs in clinical trials: lilly experience. *Biopharmaceutical Report* **3:1** 1–11, with discussion.

FLOURNOY, N., HAINES, L. M., AND ROSENBERGER, W. F. (2013). A graphical comparison of response-adaptive randomization procedures. *Statistics in Biopharmaceutical Research* **5** 126–141.

HU, F. AND ROSENBERGER, W. F. (2006). *The Theory of Response- Adaptive Randomization in Clinical Trials*. John Wiley & Sons, Inc., New York.

HU, F. AND ZHANG, L.-X. (2004). Asymptotic properties of doubly adaptive biased coin designs for multi-treatment clinical trials. *Annals of Statistics* **32** 268–301.

O'ROURKE, P. P., CRONE, R. K., VACANTI, J. P., WARE, J. H., LILLEHEI, C. W., PARAD, R. B., AND EPSTEIN, M. F. (1989). Extracorporeal membrane oxygenation and conventional medical therapy in neonates with persistent pulmonary hypertension of the newborn: a prospective randomized study. *Pediatrics* **84** 957–963.

ROSENBERGER, W. F. (1996). New directions in adaptive designs. *Statistical Science* **11** 137–149.

ROSENBERGER, W. F. (1999). Randomized play-the-winner clinical trials: review and recommendations. *Controlled Clinical Trials* **20** 328–342.

ROSENBERGER, W. F. AND HU, F. (1999). Bootstrap methods for adaptive designs. *Statistics in Medicine* **18** 1757–1767.

ROSENBERGER, W. F. AND LACHIN, J. M. (1993). The use of response-adaptive designs in clinical trials. *Controlled Clinical Trials* **14** 471–484.

ROSENBERGER, W. F. AND SESHAIYER, P. (1997). Adaptive survival trials. *Journal of Biopharmaceutical Statistics* **7** 617–624.

ROUT, C. C., ROCKE, D. A., LEVIN, J., GOUWS, E., AND REDDY, D. (1993). A reevaluation of the role of crystalloid preload in the prevention of hypotension associated with spinal anesthesia for elective Cesarean section. *Anesthesiology* **79** 262–269.

ROYALL, R. M. (1991). Ethics and statistics in randomized clinical trials. *Statistical Science* **6** 52–62, with discussion.

SIMON, R. (1991). A decade of progress in statistical methodology for clinical trials. *Statistics in Medicine* **10** 1789–1817.

TAMURA, R. N., FARIES, D. E., ANDERSEN, J. S., AND HEILIGENSTEIN, J. H. (1994). A case study of an adaptive clinical trial in the treatment of out-patients with depressive disorder. *Journal of the American Statistical Association* **89** 768–776.

TYMOFYEYEV, Y., ROSENBERGER, W. F., AND HU, F. (2007). Implementing optimal allocation in sequential binary response experiments. *Journal of the American Statistical Association* **102** 224–234.

UK Collaborative ECMO Trial Group. (1996). Collaborative randomized trial of neonatal extracorporeal membrane oxygenation. *Lancet* **348** 75–82.

WARE, J. H. (1989). Investigating therapies of potentially great benefit: ECMO. *Statistical Science* **4** 298–340, with discussion.

WEINSTEIN, M. C. (1974). Allocation of subjects in medical experiments. *New England Journal of Medicine* **291** 1278–1285.

ZHANG, L. AND ROSENBERGER, W. F. (2006). Response-adaptive randomization for clinical trials with continuous outcomes. *Biometrics* **62** 562–569.

ZHANG, L. AND ROSENBERGER, W. F. (2007). Response-adaptive randomization for survival trials: the parametric approach. *Journal of the Royal Statistical Society C* **53** 153–165.

Author Index

Randomization in Clinical Trials: Theory and Practice, Second Edition.
William F. Rosenberger and John M. Lachin.
© 2016 John Wiley & Sons, Inc. Published 2016 by John Wiley & Sons, Inc.

Subject Index

Randomization in Clinical Trials: Theory and Practice, Second Edition.
William F. Rosenberger and John M. Lachin.
© 2016 John Wiley & Sons, Inc. Published 2016 by John Wiley & Sons, Inc.

WILEY SERIES IN PROBABILITY AND STATISTICS

ESTABLISHED BY WALTER A. SHEWHART AND SAMUEL S. WILKS

Editors: *David J. Balding, Noel A. C. Cressie, Garrett M. Fitzmaurice, Geof H. Givens, Harvey Goldstein, Geert Molenberghs, David W. Scott, Adrian F. M. Smith, Ruey S. Tsay, Sanford Weisberg*
Editors Emeriti: *J. Stuart Hunter, Iain M. Johnstone, Joseph B. Kadane, Jozef L. Teugels*

The *Wiley Series in Probability and Statistics* is well established and authoritative. It covers many topics of current research interest in both pure and applied statistics and probability theory. Written by leading statisticians and institutions, the titles span both state-of-the-art developments in the field and classical methods.

Reflecting the wide range of current research in statistics, the series encompasses applied, methodological and theoretical statistics, ranging from applications and new techniques made possible by advances in computerized practice to rigorous treatment of theoretical approaches.

This series provides essential and invaluable reading for all statisticians, whether in aca-demia, industry, government, or research.

* BOX and TIAO · Bayesian Inference in Statistical Analysis

BOX · Improving Almost Anything, Revised Edition

* BOX and DRAPER · Evolutionary Operation: A Statistical Method for Process Improvement

BOX and DRAPER · Response Surfaces, Mixtures, and Ridge Analyses, Second Edition

BOX, HUNTER, and HUNTER · Statistics for Experimenters: Design, Innovation, and Discovery, Second Editon

BOX, JENKINS, REINSEL, and LJUNG · Time Series Analysis: Forecasting and Control, Fifth Edition

BOX, LUCEÑO, and Paniagua-QuiÑones · Statistical Control by Monitoring and Adjustment, Second Edition

* BROWN and HOLLANDER · Statistics: A Biomedical Introduction

CAIROLI and DALANG · Sequential Stochastic Optimization

CASTILLO, HADI, BALAKRISHNAN, and SARABIA · Extreme Value and Related Models with Applications in Engineering and Science

CHAN · Time Series: Applications to Finance with R and S-Plus®, Second Edition

CHARALAMBIDES · Combinatorial Methods in Discrete Distributions

CHATTERJEE and HADI · Regression Analysis by Example, Fourth Edition

CHATTERJEE and HADI · Sensitivity Analysis in Linear Regression

Chen · The Fitness of Information: Quantitative Assessments of Critical Evidence

CHERNICK · Bootstrap Methods: A Guide for Practitioners and Researchers, Second Edition

CHERNICK and FRIIS · Introductory Biostatistics for the Health Sciences

CHILÈS and DELFINER · Geostatistics: Modeling Spatial Uncertainty, Second Edition

CHIU, STOYAN, KENDALL and MECKE · Stochastic Geometry and Its Applications, Third Edition

CHOW and LIU · Design and Analysis of Clinical Trials: Concepts and Methodologies, Third Edition

CLARKE · Linear Models: The Theory and Application of Analysis of Variance

CLARKE and DISNEY · Probability and Random Processes: A First Course with Applications, Second Edition

* COCHRAN and COX · Experimental Designs, Second Edition

COLLINS and LANZA · Latent Class and Latent Transition Analysis: With Applications in the Social, Behavioral, and Health Sciences

CONGDON · Applied Bayesian Modelling, Second Edition

CONGDON · Bayesian Models for Categorical Data

CONGDON · Bayesian Statistical Modelling, Second Edition

CONOVER · Practical Nonparametric Statistics, Third Edition

COOK · Regression Graphics

COOK and WEISBERG · An Introduction to Regression Graphics

COOK and WEISBERG · Applied Regression Including Computing and Graphics

CORNELL · A Primer on Experiments with Mixtures

CORNELL · Experiments with Mixtures, Designs, Models, and the Analysis of Mixture Data, Third Edition

Schoutens · Levy Processes in Finance: Pricing Financial Derivatives

SCOTT · Multivariate Density Estimation

SCOTT · Multivariate Density Estimation: Theory, Practice, and Visualization

* SEARLE · Linear Models

† SEARLE · Linear Models for Unbalanced Data

† SEARLE · Matrix Algebra Useful for Statistics

† SEARLE, CASELLA, and McCULLOCH · Variance Components

SEARLE and WILLETT · Matrix Algebra for Applied Economics

SEBER · A Matrix Handbook For Statisticians

† SEBER · Multivariate Observations

SEBER and LEE · Linear Regression Analysis, Second Edition

† SEBER and WILD · Nonlinear Regression

SENNOTT · Stochastic Dynamic Programming and the Control of Queueing Systems

* SERFLING · Approximation Theorems of Mathematical Statistics

SHAFER and VOVK · Probability and Finance: It's Only a Game!

SHERMAN · Spatial Statistics and Spatio-Temporal Data: Covariance Functions and Directional Properties

SILVAPULLE and SEN · Constrained Statistical Inference: Inequality, Order, and Shape Restrictions

SINGPURWALLA · Reliability and Risk: A Bayesian Perspective

SMALL and McLEISH · Hilbert Space Methods in Probability and Statistical Inference

SRIVASTAVA · Methods of Multivariate Statistics

STAPLETON · Linear Statistical Models, Second Edition

STAPLETON · Models for Probability and Statistical Inference: Theory and Applications

STAUDTE and SHEATHER · Robust Estimation and Testing

Stoyan · Counterexamples in Probability, Second Edition

STOYAN and STOYAN · Fractals, Random Shapes and Point Fields: Methods of Geometrical Statistics

STREET and BURGESS · The Construction of Optimal Stated Choice Experiments: Theory and Methods

STYAN · The Collected Papers of T. W. Anderson: 1943–1985

SUTTON, ABRAMS, JONES, SHELDON, and SONG · Methods for Meta-Analysis in Medical Research

TAKEZAWA · Introduction to Nonparametric Regression

TAMHANE · Statistical Analysis of Designed Experiments: Theory and Applications

TANAKA · Time Series Analysis: Nonstationary and Noninvertible Distribution Theory

THOMPSON · Empirical Model Building: Data, Models, and Reality, Second Edition

THOMPSON · Sampling, Third Edition

THOMPSON · Simulation: A Modeler's Approach

THOMPSON and SEBER · Adaptive Sampling

THOMPSON, WILLIAMS, and FINDLAY · Models for Investors in Real World Markets

TIERNEY · LISP-STAT: An Object-Oriented Environment for Statistical Computing and Dynamic Graphics

TROFFAES and DE COOMAN · Lower Previsions

TSAY · Analysis of Financial Time Series, Third Edition

TSAY · An Introduction to Analysis of Financial Data with R

TSAY · Multivariate Time Series Analysis: With R and Financial Applications

UPTON and FINGLETON · Spatial Data Analysis by Example, Volume II: Categorical and Directional Data

† VAN BELLE · Statistical Rules of Thumb, Second Edition

VAN BELLE, FISHER, HEAGERTY, and LUMLEY · Biostatistics: A Methodology for the Health Sciences, Second Edition

VESTRUP · The Theory of Measures and Integration

VIDAKOVIC · Statistical Modeling by Wavelets

Viertl · Statistical Methods for Fuzzy Data

VINOD and REAGLE · Preparing for the Worst: Incorporating Downside Risk in Stock Market Investments

WALLER and GOTWAY · Applied Spatial Statistics for Public Health Data

WEISBERG · Applied Linear Regression, Fourth Edition

WEISBERG · Bias and Causation: Models and Judgment for Valid Comparisons

WELSH · Aspects of Statistical Inference

WESTFALL and YOUNG · Resampling-Based Multiple Testing: Examples and Methods for p-Value Adjustment

* WHITTAKER · Graphical Models in Applied Multivariate Statistics

WINKER · Optimization Heuristics in Economics: Applications of Threshold Accepting

WOODWORTH · Biostatistics: A Bayesian Introduction

WOOLSON and CLARKE · Statistical Methods for the Analysis of Biomedical Data, Second Edition

WU and HAMADA · Experiments: Planning, Analysis, and Parameter Design Optimization, Second Edition

WU and ZHANG · Nonparametric Regression Methods for Longitudinal Data Analysis

Yakir · Extremes in Random Fields

YIN · Clinical Trial Design: Bayesian and Frequentist Adaptive Methods

YOUNG, VALERO-MORA, and FRIENDLY · Visual Statistics: Seeing Data with Dynamic Interactive Graphics

ZACKS · Examples and Problems in Mathematical Statistics

ZACKS · Stage-Wise Adaptive Designs

* ZELLNER · An Introduction to Bayesian Inference in Econometrics

ZELTERMAN · Discrete Distributions—Applications in the Health Sciences

ZHOU, OBUCHOWSKI, and McCLISH · Statistical Methods in Diagnostic Medicine, Second Edition

Printed and bound by CPI Group (UK) Ltd, Croydon, CR0 4YY

16/04/2025

14658370-0001